FEDERAL COORDINATOR
FOR
METEOROLOGICAL SERVICES AND
SUPPORTING RESEARCH

8455 COLESVILLE ROAD, SUITE 1500
SILVER SPRING, MARYLAND 20910
301-427-2002
www.ofcm.gov

FEDERAL METEOROLOGICAL HANDBOOK
NUMBER 1

SURFACE WEATHER OBSERVATIONS AND REPORTS

FCM-H1-2005
Washington, D.C.
September 2005

CHANGE AND REVIEW LOG

Use this page to record changes, notices and reviews.

Change Number	Page Numbers	Date Posted	Initials
1	(See Change Letter)	Nov 5, 1998	BKT
2	(See Change Letter)	Sep 1, 2005	Mary Cairns
3			
4			
5			
6			
7			
8			
9			
10			

Changes are indicated by a vertical line in the margin next to the change.

Review Date	Comments	Initials

TABLE OF CONTENTS

10   TEMPERATURE AND DEW POINT

11   PRESSURE

LIST OF TABLES   Page

CHAPTER 1

INTRODUCTION

1.1 Purpose

Federal Meteorological Handbook No.1, *Surface Weather Observations and Reports* (FMH-1) defines the observing, reporting, and coding standards for surface based meteorological reports. These standards are applicable to all Federal agency programs. These standards do not inhibit agencies from doing more than is specified.

1.2 Applicability of Standards

Standards described in this Handbook are applicable only if a station has the capability to comply. The phrase "at designated stations" refers to observing stations instructed by their responsible agency to perform a specified task. These specified tasks shall be performed in accordance with the standards described in this Handbook.

This Handbook is applicable to stations taking the following types of observations:

   a. Automated - automated surface weather observing systems that prepare the meteorological reports for transmission without a certified weather observer.

   b. Augmented - automated surface weather observing systems that prepare the meteorological reports for transmission with certified weather observers signed-on to the system to add information to the observation.

   c. Manual - certified weather observers are responsible for the meteorological observations.

1.3 Relation to Other Handbooks and Manuals

Individual agencies shall issue their own manuals defining their observing procedures which implement the FMH-1 standards. Such manuals shall complement, not change, the standards contained in FMH-1. Most observing standards described in this Handbook also apply to synoptic surface observations which have coding and reporting procedures described in Federal Meteorological Handbook No.2, *Surface Synoptic Codes* (FMH-2). FMH-1 is consistent with agreements and publications of the World Meteorological Organization (WMO), the International Civil Aviation Organization (ICAO), specifically WMO No. 306 - *Manual on Codes*, and ICAO Annex 3 - *Meteorological Services for International Air Navigation,* and civil as well as military weather services.

1.4 Format of This Handbook

Chapter 1 presents an introductory overview of surface weather observations.

Chapter 2 discusses the surface weather observation program, types of observations, criteria for specials, general observing standards, and dissemination.

Chapter 3 presents the certification and quality control standards. It discusses certification of observers and quality control performed on-site, and at remote locations.

Chapter 4 defines the requirements for maintaining records of surface weather data. It discusses the types of records, preparation and maintenance of the Station Information File, types of storage, and procedures for maintaining records of surface weather data.

Chapters 5 through 11 focus on a specific element (e.g., Visibility is an element) and all the associated parameters (e.g., Prevailing Visibility and Sector Visibility are visibility parameters) of that element that appear in the weather report. The elements are: Wind (Chapter 5), Visibility (Chapter 6), Runway Visual Range (Chapter 7), Present Weather (Chapter 8), Sky Condition (Chapter 9), Temperature and Dew Point (Chapter 10), and Pressure (Chapter 11). Each chapter contains a similar format: a brief overview of the chapter, a section that describes parameters, and a section that defines further the observing and reporting standards for the reports. Each chapter also contains a Summary Table that offers an abridged listing of standards described in the chapter. By design, these summary tables do not contain all of the details found in the text. Therefore, these tables should only be used as an overview of the standards contained in the chapter.

Chapter 12 defines the coding procedures for reports in the METAR/SPECI format.

Appendix A is a Glossary.

Appendix B is a list of Abbreviations and Acronyms.

Appendix C lists Sensor Standards.

Appendix D contains Runway Visual Range Tables.

Throughout this Handbook, the following definitions apply:

    a. "shall" indicates a standard is mandatory.
    b. "should" indicates a standard is recommended.
    c. "may" indicates a standard is optional.
    d. "will" indicates futurity; it is not a requirement to be applied to standards.

1.5 Changes to The Handbook

Changes, additions, deletions, and corrections will be issued, as necessary. These changes shall be issued only by the Office of the Federal Coordinator for Meteorology (OFCM) after consultation and coordination with the Working Group for Atmospheric Observing Systems (WG/AOS).

1.6 Agency Procedures and Procedural Changes

Agencies shall issue manuals and directives to provide more detailed instructions and training to their personnel and users (see paragraph 1.3). Copies of these procedures shall be provided to the reference library (see paragraph 1.7). Agencies may also issue changes to their procedures as follows:

    a. Routine procedural changes that are in conformance with FMH-1 may be issued by an agency at any time without coordination with the WG/AOS.

    b. Procedural changes not in conformance with FMH-1 shall be coordinated with the WG/AOS when time allows. However, when necessary to meet urgent requirements, procedural changes not in conformance with FMH-1 may be issued by an agency without coordination with the WG/AOS. Copies of changes shall be forwarded to the WG/AOS for coordination and appropriate action.

1.7 Reference Library

The WG/AOS shall maintain a record and library of agency procedures, changes, and supplements issued by all participating agencies. The WG/AOS shall establish a procedure for making these procedures, changes, and supplements available to users.

1.8 Unforeseen Requirements

No set of instructions can cover all possibilities in weather observing. Observers must use their own judgment, adhering as closely as possible to this Handbook and agency operating procedures, to describe phenomena not adequately covered by specific instructions. If the observer believes Handbook procedures require change or clarification, suggested changes should be sent through normal administrative channels to the headquarters of the appropriate organization.

1.9 Other Questions and Suggestions Regarding FMH-1

Questions or suggestions about the content or organization of this Handbook should be directed to:

Office of the Federal Coordinator for Meteorology
8455 Colesville Road, Suite 1500
Silver Spring, MD 20910

Telephone:	Commercial	(301) 427-2002
	FAX	(301) 427-2007
	DSN	851-1460

CHAPTER 2

SURFACE WEATHER OBSERVATION PROGRAM

2.1 General

Surface weather observations are fundamental to all meteorological services. Observations are the basic information upon which forecasts and warnings are made in support of a wide range of weather sensitive activities within the public and private sectors.

2.2 Scope

This chapter briefly describes the Federal Government's surface weather observation program and outlines the observing program and procedures which apply to each of the agencies involved in surface weather observing. In addition, the chapter addresses the types of dissemination and the general requirements for verifying and making corrections to disseminated reports.

2.3 Surface Weather Observation Program

The Departments of Commerce (DOC), Defense (DOD), and Transportation (DOT) have established networks of stations that collectively provide the meteorological data used by the public and private sectors. As the Nation's primary civil meteorological agency, the DOC's National Weather Service (NWS) has the responsibility for observing, analyzing and forecasting weather conditions. DOD organizational elements within the U. S. Air Force (Major Commands), Marine Corps, and Navy (Naval Meteorology and Oceanography Command (NAVMETOCCOM)) take weather observations to support DOD operations worldwide. In addition to taking observations, the DOT's Federal Aviation Administration (FAA), as the agency responsible for safe operation of aircraft and efficient use of the Nation's airspace system, has the role of establishing requirements for, and disseminating aviation weather data to airspace users.

In addition to the observations taken by the above Federal agencies, observations are taken by commercial airline companies, private individuals, and local and state government agencies. These non-Federal locations are established and operated under the guidance of the NWS, in cooperation with the FAA.

2.4 Aviation Weather Observing Locations

There shall not be more than one official observation for a specific location at any one time. For meteorological observations, the observing location is defined as the point or points at which the various elements of the observation are evaluated. At a large airport, the locations may be defined as follows:

a. For clouds, surface visibility, and weather, the observing location may be at the touchdown area of the primary runway.

b. For tower visibility, the observing location shall be the Airport Traffic Control Tower (ATCT).

c. For temperature, dew point, and wind, the observing location may be the center of the runway complex.

d. For the location, type, and frequency of lightning (see paragraph 12.7.1.j(2)), the observing location may be the Airport Location Point (ALP) .

Specific details on the siting of observing equipment can be found in the Station Information File (see paragraph 4.3.1).

Manual and augmented weather observations may also contain information on phenomena occurring at other than the station. In these cases, the point(s) where the phenomenon occurs is not considered to be an additional observing location.

2.5 Types of Reports

2.5.1 Aviation Routine Weather Report (METAR)

METAR is the primary observation code used in the United States to satisfy requirements for reporting surface meteorological data. METAR contains a report of wind, visibility, runway visual range, present weather, sky condition, temperature, dew point, and altimeter setting collectively referred to as "the body of the report". In addition, coded and/or plain language information which elaborates on data in the body of the report may be appended to the METAR. This significant information can be found in the section referred to as "Remarks". The contents of the remarks will vary according to the type of weather station. At designated stations, the METAR may be abridged to include one or more of the above elements.

2.5.2 Aviation Selected Special Weather Report (SPECI)

SPECI is an unscheduled report taken when any of the criteria given in paragraph 2.5.2.a have been observed. SPECI shall contain all data elements found in a METAR plus additional plain language information which elaborates on data in the body of the report. All SPECIs shall be made as soon as possible after the relevant criteria are observed.

a. **Criteria for SPECI**

   (1) WIND SHIFT. Wind direction changes by 45 degrees or more in less than 15 minutes and the wind speed is 10 knots or more throughout the wind shift.

   (2) VISIBILITY. Surface visibility as reported in the body of the report decreases to less than, or if below, increases to equal or exceed:

   (a) 3 miles.
   (b) 2 miles.
   (c) 1 mile.
   (d) The lowest standard instrument approach procedure minimum as published in the National Ocean Service (NOS) *U.S. Terminal Procedures*. If none published, use 1/2 mile.

   (3) RUNWAY VISUAL RANGE (RVR). The highest value from the designated RVR runway decreases to less than, or if below, increases to equal or exceed 2,400 feet during the preceding 10 minutes. U.S. military stations may not report a SPECI based on RVR.

   (4) TORNADO, FUNNEL CLOUD, OR WATERSPOUT.

   (a) is observed.
   (b) disappears from sight, or ends.

   (5) THUNDERSTORM.

   (a) begins (a SPECI is not required to report the beginning of a new thunderstorm if one is currently reported).
   (b) ends.

(6) PRECIPITATION.

  (a) hail begins or ends.
  (b) freezing precipitation begins, ends, or changes intensity.
  (c) ice pellets begin, end, or change intensity.

(7) SQUALLS. When squalls occur.

(8) CEILING. The ceiling (rounded off to reportable values) forms or dissipates below, decreases to less than, or if below, increases to equal or exceed:

  (a) 3,000 feet.
  (b) 1,500 feet.
  (c) 1,000 feet.
  (d) 500   feet.
  (e) The lowest standard instrument approach procedure minimum as published in the National Ocean Service (NOS) *U.S. Terminal Procedures*. If none published, use 200 feet.

(9) SKY CONDITION. A layer of clouds or obscurations aloft is present below 1,000 feet and no layer aloft was reported below 1,000 feet in the preceding METAR or SPECI.

(10) VOLCANIC ERUPTION. When an eruption is first noted.

(11) AIRCRAFT MISHAP. Upon notification of an Aircraft Mishap unless there has been an intervening observation.

(12) MISCELLANEOUS. Any other meteorological situation designated by the responsible agency, or which, in the opinion of the observer, is critical.

  b. The SPECI criteria are only applicable to stations that have the capability of evaluating the event. For example, visually evaluated elements, such as a tornado, are not applicable to non-staffed automated stations.

2.6 Observing Standards Applicable to All Stations

2.6.1 Use of Certified Observers. All personnel performing an observation function shall be certified in accordance with paragraph 3.3.1. Certification may be limited in accordance with observer responsibilities.

2.6.2 Backup. Backup refers to a method, in accordance with agency procedures, for providing meteorological reports, parts of reports, documentation, or communication of reports when the primary method is unavailable.

2.6.3 Rounding Figures. Except where otherwise designated in this Handbook, the rounding of numbers shall be accomplished as follows: If the fractional part of a positive number to be dropped is equal to or greater than one-half, the preceding digit shall be increased by one. If the fractional part of a negative number to be dropped is greater than one-half, the preceding digit shall be decreased by one. In all other cases, the preceding digit shall remain unchanged. For example, 1.5 becomes 2, -1.5 becomes -1, 1.3 becomes 1, and -2.6 becomes -3.

2.6.4 Time Used in Reports. With the exception of designated stations which shall transmit reports in accordance with agency instructions, METAR shall be transmitted at fixed intervals with SPECI transmitted when any of the criteria in paragraph 2.5.2.a occurs or is noted. Each station's schedule for transmitting reports shall be included in the Station Information File (see paragraph 4.3.1).

a. **Accuracy of Time in Reports.** A procedure shall be established to assure that the accuracy of the timing device used to establish times in the observation program are within ±1 minute of the U.S. Naval Observatory time.

b. **Scheduled Time of Report.** The scheduled time of the METAR shall be the Coordinated Universal Time (UTC) a METAR is required to be available for transmission.

c. **Actual Date and Time of Observation.** The actual date and time of METAR shall be the time the last element of the observation was observed. The actual time of a SPECI shall be the time the criteria for the SPECI was met or noted.

d. **Time Disseminated in Observations.** All times disseminated in observations shall reference the 24-hour UTC clock, e.g., 1:47 A.M. shall be referred to as 0147 and 1:47 P.M. as 1347. The times 0000 and 2359 shall indicate the beginning and ending of the day, respectively.

e. **Date and Time Entered in Observations.** All dates and times entered in observations shall be with reference to the 24-hour clock. The times that are disseminated as part of the observation shall be entered in UTC. However, at the discretion of the responsible agency, those times used to otherwise document the observation or other related observational data may be either Local Standard Time (LST) or UTC. The time standard selected shall be clearly indicated on all records; if LST is used, the number of hours used to convert to UTC shall also be indicated.

2.6.5   Sensor Siting Standards. All installations of sensors shall be in accordance with the latest *Federal Standard for Siting Meteorological Sensors at Airports* published by the OFCM. Presently installed sensors may be operated at their present location. However, if they must be relocated, the Federal standards shall be followed.

2.6.6   Algorithms Used by Automated Stations. Automated stations shall use algorithms that conform with the latest *Federal Standard Algorithms for Automated Weather Observing Systems* published by the OFCM. These algorithms do not apply to previously authorized systems, which may continue to operate until replaced or modified.

2.7   Recency of Observed Elements

2.7.1   Recency of Observed Elements at Automated Stations. Individual elements entered in an observation shall, as closely as possible, reflect conditions existing at the actual time of observation. For those elements that the human observer evaluates using spatial averaging techniques (e.g., sky cover and visibility), the automated station substitutes time averaging of sensor data. Therefore, in an automated observation, sky condition shall be an evaluation of sensor data gathered during the 30-minute period ending at the actual time of the observation. All other elements evaluated shall be based on sensor data that is within 10 minutes or less of the actual time of the observation.

2.7.2   Recency of Observed Elements at Manual Stations. Individual elements entered in an observation shall, as closely as possible, reflect conditions existing at the actual time of observation. Elements entered shall have been observed within 15 minutes of the actual time of the observation. Gusts and squalls shall be reported if observed within 10 minutes of the actual time of the observation. Observation of elements shall be made as close to the scheduled time of the observation as possible to meet filing deadlines, but in no case shall these observations be started more than 15 minutes before the scheduled time.

2.8 Dissemination

For purposes of this Handbook, dissemination is the act of delivering a completed report to users.

2.8.1 Types of Dissemination. There are two general types of dissemination:

a. **Local** -- The transmission or delivery of a weather report to individual or groups of users in the service area of the weather station.

b. **Long-line** -- The transmission of a weather report beyond the service area of the weather station.

2.8.2 Dissemination Requirements. All reports shall be given local dissemination. At designated stations, reports shall be given long-line dissemination. When reports are corrected, the corrected reports shall be given the same dissemination as the reports being corrected.

2.8.3 Dissemination Priority. If reports cannot be disseminated simultaneously, local and long-line, they should be disseminated first to the local airport traffic control users. Further dissemination priorities shall be defined by the responsible agencies.

2.8.4 Corrections to Transmitted Data. Corrections shall be disseminated, as soon as possible, whenever an error is detected in a transmitted report. However, if the erroneous data has been superseded by a later report (with the same or more complete dissemination), it shall not be necessary to transmit the corrected report. Corrections transmitted shall consist of the entire corrected report. The original date and time of the report shall be used as the date and time in the corrected report.

2.9 Report Filing Time

SPECIs shall be completed and transmitted, as soon as possible. Agencies shall establish filing deadlines for all METARs; the filing deadlines shall be no sooner than necessary to assure the availability of the report at its scheduled time. METARs shall not be transmitted sooner than 10 minutes before their scheduled time.

2.10 Delayed Reports

When transmission of an observation is delayed until time for the next regularly scheduled report, only the latest report shall be transmitted. In the record of observations, the remark **FIBI** (Filed But Impractical to Transmit) shall be appended in parentheses to the report that was not transmitted to indicate the report was not transmitted. The remark FIBI shall not be included in any local dissemination of the report.

When a SPECI is not transmitted long-line, the later SPECI shall be transmitted long-line only when the overall change between the last transmitted report and the current report satisfies the criteria for a SPECI. If the SPECI is not transmitted long-line, the remark FIBI shall be appended to the report in the record of observations. The SPECI shall be disseminated locally.

Reports of Volcanic Eruption shall be disseminated, by any means possible, regardless of the delay.

CHAPTER 3

CERTIFICATION AND QUALITY CONTROL

3.1 General

In order to ensure validity and reliability of the Federal Government's meteorological data, the use of standards, certification, and quality control must be an inherent program element. Without these essential elements, the credibility of the Nation's climatological database would become suspect and invoke mistrust.

3.2 Scope

This chapter prescribes the general standards for certification of observers and quality control of weather observations. It separately addresses the standards and procedures applicable to all stations. It also discusses the requirement for quality control at National Centers.

3.3 Certification Requirements

3.3.1 Certification of Observers. Agencies shall have an observer certification program, and only observers certified through that program shall be authorized to take weather reports. The NWS shall have the responsibility for certifying observers at all civil stations.

The certification of the observer shall attest to the fact that in the view of the certifying agency the observer has:

    a. **Acceptable vision.** The observer shall have distant vision of at least 20/30 (Snellen) in at least one eye, corrected if necessary.

    b. **Adequate Training.** The level of training would be commensurate with the level of the weather reporting function to be undertaken by the candidate.

    c. **Demonstrated Ability to Take Required Weather Reports.** This demonstrated ability would be dependent on the level of the weather reporting function to be undertaken by the candidate.

3.3.2 Certification of Stations. All stations where weather reports are taken shall be approved by the responsible agency. The procedures for the approval shall also be established by that agency. The procedures shall include verification that:

    a. the instruments to be used meet the minimum standards for accuracy, range, and resolution for weather elements as prescribed in this document (Appendix C);

    b. the installation of the automated sensors satisfy the siting requirements prescribed in the *Federal Standard for Siting Meteorological Sensors at Airports*;

    c. the algorithms used in automated sensors are in accordance with the algorithms prescribed in the *Federal Standard Algorithms for Automated Weather Observing Systems used for Aviation Purposes*;

    d. the proposed maintenance program to support the operation of the station is acceptable to the responsible agency; and

    e. all observers are certified at a level commensurate with their duties in accordance with this Handbook.

3.4 Quality Control

3.4.1 Quality Control of Observing Programs. Each agency shall establish a quality control program to ensure that their surface weather reporting stations utilize proper procedures. The primary objective of the quality control program shall be to ascertain that:

a. the siting and exposure of instruments is the best practical, and are still within acceptable limits;

b. instruments are in good order and have been compared to the standard sensors, as required;

c. standard procedures are being used to generate weather reports at the station;

d. the observation program at the station satisfies the requirements for weather reports at that location; and

e. any observers taking weather reports are certified.

3.4.2 Quality Control of Instruments and Sensors

a. **Comparison of Portable Transfer-Standard Sensors.** Agencies shall establish a procedure to routinely compare portable transfer-standard instruments and sensors, which are used during station inspections, to standards that ensure compliance with the accuracy requirements, as listed in Appendix C, of this Handbook.

b. **Comparison of Instruments and Sensors.** Agencies shall establish a procedure to periodically compare the instruments and sensors used at operational weather reporting stations to standards that ensure compliance with the accuracy requirements, as listed in Appendix C, of this Handbook.

c. **Calibration and Standardization of Sensors.** Agencies shall establish procedures to calibrate and standardize sensors. Calibration and standardization should be performed at least annually, after installation, and after any major maintenance is performed on a sensor.

d. **Routine Maintenance.** Each agency shall establish a schedule of maintenance for equipment at stations. Unless relieved of the responsibility, observers at stations shall determine the operational acceptability of meteorological equipment consistent with agency policy. In addition to having the responsibility for the operational status of meteorological equipment, the observer shall also have the final authority for deciding the operational status of any meteorological equipment used in the weather reporting program.

3.4.3 Quality Control of Weather Reports. Each agency shall establish a near real-time quality control program for all stations. This program shall be used to quickly detect repetitious errors being made by observers at the station. The program may use station personnel or personnel at another location with access to the records and reports made at the station.

a. **Pre-Dissemination Quality Control.** The quality control performed at surface weather reporting stations prior to any dissemination of the weather report is the most important of all quality control operations. Once an erroneous report has been given to users, it is impossible to ensure that corrections are received by that same group of users. Therefore, all operational weather reporting stations shall have as high a level of pre-dissemination quality control procedures as practicable. This check should consist of recalculating computed data, verifying the syntax of the recorded weather reports, and comparing the recorded weather report against the reports recorded on any local dissemination devices.

b. **Post-Dissemination Quality Control.** All surface weather reports shall be checked at the site for errors after dissemination and prior to the next weather report. If possible, the disseminated report shall be compared with the original report to verify that no errors were generated during the dissemination process.

3.5 Quality Control Performed at Central Locations

Insofar as possible, all agencies should implement quality control checking at a central location on a timely basis. Agencies shall devise a method to provide feedback to the observer in the case of errors detected in manual weather reports.

3.6 Customer Feedback

Agencies shall encourage customers to comment on the performance of their observing programs. In this context, customers are anyone utilizing the data from the programs.

It is recommended that all public issuances of information on the observing systems include an address of the office designated to process customer feedback.

CHAPTER 4

OBSERVATIONAL RECORDS

4.1 General

All meteorological data collected as part of the surface weather observation program are valuable information. Consequently, these data require retention, storage, and archival.

4.2 Scope

This chapter prescribes the minimum standards for maintaining a record of the operational status of weather reporting stations and maintaining records of surface weather reports. Because of the lasting value of all the data collected in the surface weather reporting program, agencies are encouraged to devise methods to store as much of the data as possible within their physical and financial constraints. Because of the generally common interest in historical and archived data, agencies shall provide records to the National Climatic Data Center (NCDC).

4.3 Types of Records

4.3.1 Station Information File. Each agency shall establish a Station Information File for each station under their jurisdiction. The file shall be a record of characteristics of each station. Table 4-1 presents a list of data required from most stations. Copies of all Station Information Files shall be available upon request from the responsible agency when needed to interpret weather records from a particular station.

Table 4-1. Content of Station Information File

Physical Characteristics		
Station Name	Latitude/Longitude	Type of Station
Airport Name	Climatological Elevation	Description of
Station Identifier	Field Elevation	Significant
WMO Index Number	Ground Elevation	Topography
Time Zone	Station Elevation	
Observation Schedule		
Types of Reports	Schedule for Reports	Hours of Operation
Observation Program		
Elements Observed	SPECI	
Long-Line Communications Circuits	Transmitted (Y/N)	
Sensor Data		
System Configuration	Location of Sensors	Non-standard sensor siting
Types of Sensors		

a. **Maintenance of Station Information File.** The Station Information File should be created on or before the date of station establishment to ensure the timely notification of the station's existence. The file shall be updated when any data in Table 4-1 or in the file changes, or the station closes, noting the date when each change is implemented. Corrections to erroneous station information shall be noted as such, along with the historical extent of the error. Agencies shall establish procedures to ensure the validity of their stations' Station Information Files. It is recommended that the content of the most recent file be reviewed as part of each station inspection.

b. **Station Information File at NCDC.** A copy of the Station Information File for each observing location shall be sent to the agency group representatives (NOAA, U.S. Navy, and U.S. Air Force) at the Federal Climate Complex, Asheville, North Carolina, and made available to the DOC/NOAA/NCDC. A copy of the file shall be sent when a station is established, when any data listed in Table 4-1 or in the file changes or is corrected, and when a station closes. This file shall become a part of that station's archived record.

4.3.2   Retention of Station Observational Records. Station observational records include regularly observed meteorological data from manual stations and automated systems. Each agency shall define procedures for retaining all observational records in accordance with agency data retention and archival schedules under general METAR/SPECI guidelines:

a. **Short-term Storage.** Storage of observational records for 4 or more days is required to assist in sensor/system maintenance, and verification of sensor/system records in the event of an aircraft mishap.

b. **Long-term Retention.** Each agency shall establish procedures to provide long-term retention of all observational records for 5 years to satisfy requirements for local studies and to support litigation.

c. **Archive.** The DOC/NOAA/NCDC is responsible for archiving observational records, as necessary, to satisfy requirements to monitor the Nation's climate. Agencies shall include procedures to ensure the delivery of observational records to the agency group representatives (NOAA, U.S. Navy, and U.S. Air Force) at the Federal Climate Complex.

4.3.3   Station System/Sensor/Configuration Records. Each agency shall maintain information about changes in the configuration of each station's systems/sensors, and in their operational status. This information may be collected as part of the Station Information File or the Station Observational Records and, as such, maintained, retained, and archived as outlined in the preceding Sections.

CHAPTER 5

WIND

5.1 General

Wind shall be measured in terms of velocity, a vector that includes direction and speed.

5.2 Scope

This chapter prescribes the standards for observing and reporting wind data.

5.3 Wind Parameters

As used in this chapter, wind is the horizontal motion of the air past a given point and includes:

a. **Direction.** The direction, in tens of degrees, from which the wind is blowing.

b. **Speed.** The rate, in knots, at which the wind passes a given point.

c. **Gusts.** The description of the variability of the wind speed.

d. **Peak wind speed.** The maximum instantaneous wind speed measured.

e. **Wind Shift.** A change in wind direction.

5.4 Wind Observing Standards

Wind direction, speed, and gusts shall be determined at all stations. All other wind related parameters shall be determined at designated stations.

5.4.1 Wind Direction. The wind direction shall be determined by averaging the direction over a 2-minute period. When the wind direction sensor(s) is out of service, at designated stations, the direction may be estimated by observing the wind cone or tee, movement of twigs, leaves, smoke, etc., or by facing into the wind in an unsheltered area.

5.4.2 Variable Wind Direction. The wind direction may be considered variable if, during the 2-minute evaluation period, the wind speed is 6 knots or less. Also, the wind direction shall be considered variable if, during the 2-minute evaluation period, it varies by 60 degrees or more when the average wind speed is greater than 6 knots.

5.4.3 Wind Speed. The wind speed shall be determined by averaging the speed over a 2-minute period. At designated stations, Table 5-1 shall be used to estimate wind speeds when instruments are out of service or the wind speed is below the starting speed of the anemometer in use.

5.4.4 Wind Gust. The wind speed data for the most recent 10 minutes shall be examined to evaluate the occurrence of gusts. Gusts are indicated by rapid fluctuations in wind speed with a variation of 10 knots or more between peaks and lulls.   The speed of a gust shall be the maximum instantaneous wind speed.

5.4.5 Peak Wind Speed. Peak wind data shall be determined with wind speed recorders. The peak wind speed shall be the maximum instantaneous speed measured since the last routine METAR.

Table 5-1. Estimating Wind Speed

Knots	Specification	Knots	Specification
<1	Calm; smoke rises vertically.	22-27	Large branches in motion; whistling heard in overhead wires; umbrellas used with difficulty.
1-3	Direction of wind shown by smoke drift not by wind vanes.	28-33	Whole trees in motion; inconvenience felt walking against wind.
4-6	Wind felt on face; leaves rustle; vanes moved by wind.	34-40	Breaks twigs off trees; impedes progress.
7-10	Leaves and small twigs in constant motion; wind extends light flag.	41-47	Slight structural damage occurs.
11-16	Raises dust, loose paper; small branches moved.	48-55	Trees uprooted; considerable damage occurs.
17-21	Small trees in leaf begin to sway; crested wavelets form on inland waters.	56-71	Widespread damage.

5.4.6   Wind Shifts. Wind data shall be examined to determine the occurrence of a wind shift. A wind shift is indicated by a change in wind direction of 45 degrees or more in less than 15 minutes with sustained winds of 10 knots or more throughout the wind shift.

5.4.7   Wind Sensor Range, Accuracy, and Resolution. The required range, accuracy, and resolution for wind sensors are listed in Appendix C.

5.5   Wind Reporting Standards

5.5.1   Units of Measure and Resolution for Wind. Wind direction and speed shall be reported in the body of all observations. Direction shall be reported in tens of degrees with reference to true north and speed shall be reported in knots (see paragraph 12.6.5).

5.5.2   Calm Winds. When no motion of air is detected, the wind shall be reported as calm (see paragraph 12.6.5.d).

5.5.3   Variable Wind Direction. When the wind direction is variable, a variable wind entry shall be reported as part of the wind group in the body of the report (see paragraphs 12.6.5.b and 12.6.5.c).

5.5.4   Wind Gust. When a gust is detected within 10 minutes of the actual time of the observation, the maximum instantaneous speed shall be reported (see paragraph 12.6.5.a).

5.5.5   Peak Wind Data. The peak wind shall be reported in the remarks section whenever the maximum instantaneous speed in knots (since the last METAR) is greater than 25 knots (see paragraph 12.7.1.d).

5.5.6   Wind Shifts. The wind shift and the time of occurrence shall be reported in the remarks section (see paragraph 12.7.1.e).

5.6 Summary of Wind Observing and Reporting Standards

Table 5-2 summarizes the wind observing and reporting standards.

Table 5-2. Summary of Wind Observing and Reporting Standards

Parameter	Observing and Reporting Standard
Wind direction	2-minute average in 10 degree increments with respect to true north is reported.
Wind speed	2-minute average speed in knots is reported.
Wind gust	The maximum instantaneous speed in knots in the past 10 minutes is reported.
Peak wind	The maximum instantaneous speed in knots (since the last scheduled report) shall be reported whenever the speed is greater than 25 knots.
Wind shifts	Wind shift and the time the shift occurred is reported.

CHAPTER 6

VISIBILITY

6.1 General

Visibility is a measure of the opacity of the atmosphere. An automated, instrumentally-derived visibility value is a sensor value converted to an appropriate visibility value using standard algorithms and is considered to be representative of the visibility in the vicinity of the airport runway complex. A manually-derived visibility value is obtained using the "prevailing visibility" concept. In this chapter, the term "prevailing visibility" shall refer to both manual and instrument derived visibility values.

6.2 Scope

This chapter describes the standards for observing and reporting visibility.

6.3 Visibility Parameters

The visibility parameters are:

a. **Prevailing visibility.** The visibility that is considered representative of visibility conditions at the station; the greatest distance that can be seen throughout at least half the horizon circle, not necessarily continuous.

b. **Sector visibility.** The visibility in a specified direction that represents at least a 45 degree arc of the horizon circle.

c. **Surface visibility.** The prevailing visibility determined from the usual point of observation.

d. **Tower visibility.** The prevailing visibility determined from the airport traffic control tower (ATCT) at stations that also report surface visibility.

6.4 Visibility Observing Standards.

Visibility may be manually determined at either the surface, the tower level, or both. If visibility observations are made from just one level, e.g., the airport traffic control tower, that level shall be considered the "usual point of observation" and that visibility shall be reported as surface visibility. If visibility observations are made from both levels, the visibility at the tower level may be reported as tower visibility.

Visibility may be automatically determined by sensors operating in accordance with the *Federal Standard Algorithms for Automated Weather Observing Systems Used for Aviation Purposes*. This visibility algorithm calculates a mean visibility which is the sensor equivalent of prevailing visibility. The visibility data during the period of observation are examined to determine if variable visibility shall be reported.

6.4.1 Observing Sites.

Where the observer's view of the horizon is obstructed, the observer shall move to as many locations as necessary and practicable within the time allotted for the observation to view as much of the horizon as possible. In this respect, natural obstructions, such as trees, hills, etc., are not obstructions to the horizon. These natural obstructions define the horizon.

For automated weather observing stations, the visibility sensor shall be located, in accordance with the *Federal Standard for Siting Meteorological Sensors at Airports*.

6.4.2 Manual Observing Aids.

Agencies shall establish procedures to ensure that insofar as possible, dark or nearly dark objects viewed against the horizon sky shall be used during the day, and unfocused lights of moderate intensity (about 25 candela) shall be used during the night as reference points for manually determining visibility. In addition, visibility sensors may be used to assist the observer in the evaluation.

6.4.3 <u>Observer Adaptation to Ambient Light Conditions</u>. Agencies shall establish procedures to ensure that observer's eyes shall be accustomed to the ambient lighting conditions before manual visibility observations are taken.

6.4.4 <u>Visibility</u>. Manually-derived visibility shall be evaluated as frequently as practicable. All available visibility reference points shall be used. The greatest distances that can be seen in all directions around the horizon circle shall be determined. When the visibility is greater than the distance to the farthest reference point, the greatest distance seen in each direction shall be estimated. This estimate shall be based on the appearance of the most distant visible reference points. If they are visible with sharp outlines and little blurring of color, the visibility is much greater than the distance to them. If they can barely be seen and identified, the visibility is about the same as the distance to them. After visibilities have been determined around the entire horizon circle, they shall be resolved into a single value for reporting purposes. To do this, the greatest distance that can be seen throughout at least half the horizon circle, not necessarily continuous shall be used; this is prevailing visibility. If the visibility is varying rapidly during the time of the observation, the average of all observed values around the horizon circle shall be used for reporting purposes.

6.4.5 <u>Variable Prevailing Visibility</u>. If the prevailing visibility rapidly increases and decreases by 1/2 statute mile or more, during the time of the observation, and the prevailing visibility is less than 3 miles, the visibility is considered to be variable.

6.4.6 <u>Sector Visibility</u>. When the manually-derived visibility is not uniform in all directions, the horizon circle shall be divided into arcs that have uniform visibility and represent at least one eighth of the horizon circle (45 degrees). The visibility that is evaluated in each sector is sector visibility.

6.5 <u>Visibility Reporting Standards</u>

6.5.1 <u>Unit of Measure</u>. Visibility shall be reported in statute miles.

6.5.2 <u>Prevailing Visibility</u>. Prevailing visibility shall be reported in all weather observations. The reportable values for visibility are listed in Table 6-1. If the actual visibility falls halfway between two reportable values, the lower value shall be reported (see paragraph 12.6.6).

6.5.3 <u>Variable Prevailing Visibility</u>. Variable prevailing visibility shall be reported if the prevailing visibility is less than 3 miles and rapidly increases or decreases by 1/2 statute mile or more during the time of observation. The minimum and maximum visibility values observed shall be reported in remarks section (see paragraph 12.7.1.g).

6.5.4 <u>Tower Visibility</u>. Tower visibility shall be reported, in accordance with agency procedures (see paragraph 12.7.1.f).

6.5.5 <u>Surface Visibility</u>. Surface visibility shall be the prevailing visibility from the surface at manual stations or the visibility derived from sensors at automated stations (see paragraph 12.7.1.f).

6.5.6 <u>Visibility At Second Location</u>. When an automated station uses a meteorological discontinuity visibility sensor, remarks shall be added to identify visibility at the second location which differ from the visibility in the body of the report (see paragraph 12.7.1.i).

6.5.7 <u>Sector Visibility</u>. Sector visibility shall be reported in remarks when it differs from the prevailing visibility by one or more reportable values and either the prevailing or sector visibility is less than 3 miles (see paragraph 12.7.1.h).

Table 6-1. Reportable Visibility Values

Source of Visibility Report							
Automated			**Manual**				
M1/4	2	9[a]	0	5/8	1 5/8	4	12
1/4	2 1/2	10	1/16	3/4	1 3/4	5	13
1/2	3		1/8	7/8	1 7/8	6	14
3/4	4		3/16	1	2	7	15
1	5		1/4	1 1/8	2 1/4	8	20
1 1/4	6[a]		5/16	1 1/4	2 1/2	9	25
1 1/2	7		3/8	1 3/8	2 3/4	10	30
1 3/4	8[a]		1/2	1 1/2	3	11	35[b]

a. These values may not be reported by some automated stations.
b. Further values in increments of 5 statute miles may be reported, i.e., 40, 45, 50, etc.

6.6 Summary of Visibility Observing and Reporting Standards

Table 6-2 summarizes the applicability of visibility standards.

Table 6-2. Summary of Visibility Observing and Reporting Standards and Procedures

Visibility	Type of Station	
	Automated	**Manual**
Surface	Represents 10-minutes of sensor outputs.	Visual evaluation of visibility around the horizon.
Variable	Reported when the prevailing visibility varies by 1/2 mile or more and the visibility is less than 3 miles.	
Tower	Augmented.	Reported at stations with an ATCT.
Sector	Not reported.	Reported at all stations.

CHAPTER 7

RUNWAY VISUAL RANGE

7.1 General

The runway visual range (RVR) is an instrumentally derived value that represents the horizontal distance a pilot may see down the runway.

7.2 Scope

This chapter describes the standards for observing and reporting RVR at designated stations.

7.3 Runway Visual Range (RVR) Parameter.

The runway visual range is the maximum distance at which the runway, or the specified lights or markers delineating it, can be seen from a position above a specified point on its center line. This value is normally determined by visibility sensors located alongside and higher than the center line of the runway. RVR is calculated from visibility, ambient light level, and runway light intensity.

7.4 Runway Visual Range Observing Standards

It is common practice to use a transmissometer or forward scatter meter as the RVR visibility sensor. A transmissometer measures the transmittance of the atmosphere over a baseline distance while a forward scatter meter measures the extinction coefficient of the atmosphere. RVR is then derived from equations that also account for ambient light (background luminance) and runway light intensity based on the expected detection sensitivity of the pilot's eye. RVR Tables are contained in Appendix D.

7.4.1 Observing Positions. The location of the RVR visibility sensor should be within 500 feet of the runway centerline and within 1,000 feet of the designated runway threshold.

7.4.2 Day-Night Observations for Transmissometers. The day scale shall be used in the evening until low intensity lights on or near the airport complex are clearly visible. The night scale shall be used in the morning until these lights begin to fade. Alternately, a day-night switch may be used to determine which scale should be used.

7.5 Runway Visual Range Reporting Standards

RVR shall be reported whenever the prevailing visibility is 1 statute mile or less and/or the RVR for the designated instrument runway is 6,000 feet or less. RVR shall be reported in the body of the METAR/SPECI report (see paragraphs 12.6.7).

7.5.1 Multiple Runway Visual Range Sensors. At automated stations where it is applicable, RVR values for as many as four designated runways can be reported for long-line dissemination (see paragraph 12.6.7). At manual stations, only RVR for the designated runway shall be reported.

7.5.2 Units of Measure. RVR is measured in increments of 100 feet up to 1,000 feet, increments of 200 feet from 1,000 feet to 3,000 feet, and increments of 500 feet above 3,000 feet to 6,000 feet.

7.5.3 Runway Visual Range Based on a Transmissometer. Ten-minute extreme values (highest and lowest) of transmittance shall be reported. Manually reported RVR shall be based on light setting 5 for either day or night time conditions, regardless of the light setting actually in use. One RVR value shall be reported if the ten-minute high and low value are the same.

7.6 Summary of Runway Visual Range Observing and Reporting Standards

Table 7-1 summarizes the RVR observing and reporting standards.

Table 7-1. Summary of RVR Observing and Reporting Standards

RVR	Observing and Reporting
Number of RVRs	Up to 4[a]
RVR Light Setting	5 for transmissometer systems
When Reported	When visibility ≤ 1 statute mile AND/OR RVR ≤ 6,000 feet

a. Manual observations shall contain only one RVR.

CHAPTER 8

PRESENT WEATHER

8.1 General

Present weather includes precipitation, obscurations, well-developed dust/sand whirls, squalls, tornadic activity, sandstorms, and duststorms. Present weather may be evaluated instrumentally, manually, or through a combination of instrumental and manual methods.

8.2 Scope

This chapter prescribes the standards for observing and reporting present weather. The types of present weather reported vary according to the type of station defined by the responsible agency.

8.3 Present Weather Parameters

8.3.1 Precipitation. Precipitation is any of the forms of water particles, whether liquid or solid, that fall from the atmosphere and reach the ground. The types of precipitation are:

a. **Drizzle.** Fairly uniform precipitation composed exclusively of fine drops with diameters of less than 0.02 inch (0.5 mm) very close together. Drizzle appears to float while following air currents, although unlike fog droplets, it falls to the ground.

b. **Rain.** Precipitation, either in the form of drops larger than 0.02 inch (0.5 mm), or smaller drops which, in contrast to drizzle, are widely separated.

c. **Snow.** Precipitation of snow crystals, mostly branched in the form of six-pointed stars.

d. **Snow Grains.** Precipitation of very small, white, and opaque grains of ice.

e. **Ice Crystals (Diamond Dust).** A fall of unbranched (snow crystals are branched) ice crystals in the form of needles, columns, or plates.

f. **Ice Pellets.** Precipitation of transparent or translucent pellets of ice, which are round or irregular, rarely conical, and which have a diameter of 0.2 inch (5 mm), or less. There are two main types:

(1) Hard grains of ice consisting of frozen raindrops, or largely melted and refrozen snowflakes.

(2) Pellets of snow encased in a thin layer of ice which have formed from the freezing, either of droplets intercepted by the pellets, or of water resulting from the partial melting of the pellets.

g. **Hail.** Precipitation in the form of small balls or other pieces of ice falling separately or frozen together in irregular lumps.

h. **Small Hail and/or Snow Pellets.** Precipitation of white, opaque grains of ice. The grains are round or sometimes conical. Diameters range from about 0.08 to 0.2 inch (2 to 5 mm).

i. **Unknown Precipitation.** Precipitation type that is reported if the automated station detects the occurrence of precipitation but the precipitation discriminator cannot recognize the type.

8.3.2 Obscurations. Any phenomenon in the atmosphere, other than precipitation, that reduces the horizontal visibility.

a. **Mist.** A visible aggregate of minute water particles suspended in the atmosphere that reduces visibility to less than 7 statute miles but greater than or equal to 5/8 statute miles.

b. **Fog.** A visible aggregate of minute water particles (droplets) which are based at the Earth's surface and reduces horizontal visibility to less than 5/8 statute mile and, unlike drizzle, it does not fall to the ground.

c. **Smoke.** A suspension in the air of small particles produced by combustion. A transition to haze may occur when smoke particles have traveled great distances (25 to 100 miles or more) and when the larger particles have settled out and the remaining particles have become widely scattered through the atmosphere.

d. **Volcanic Ash.** Fine particles of rock powder that originate from a volcano and that may remain suspended in the atmosphere for long periods.

e. **Widespread Dust.** Fine particles of earth or other matter raised or suspended in the air by the wind that may have occurred at or far away from the station which may restrict horizontal visibility.

f. **Sand.** Sand particles raised by the wind to a height sufficient to reduce horizontal visibility.

g. **Haze.** A suspension in the air of extremely small, dry particles invisible to the naked eye and sufficiently numerous to give the air an opalescent appearance.

h. **Spray.** An ensemble of water droplets torn by the wind from the surface of an extensive body of water, generally from the crests of waves, and carried up a short distance into the air.

8.3.3 Other Weather Phenomena

a. **Well-developed Dust/Sand Whirl.** An ensemble of particles of dust or sand, sometimes accompanied by small litter, raised from the ground in the form of a whirling column of varying height with a small diameter and an approximately vertical axis.

b. **Squall.** A strong wind characterized by a sudden onset in which the wind speed increases at least 16 knots and is sustained at 22 knots or more for at least one minute (see paragraph 12.6.8.e.(1)).

c. **Funnel Cloud (Tornadic Activity)**

   (1) **Tornado.** A violent, rotating column of air touching the ground.

   (2) **Funnel Cloud.** A violent, rotating column of air which does not touch the surface.

   (3) **Waterspout.** A violent, rotating column of air that forms over a body of water, and touches the water surface.

d. **Sandstorm.** Particles of sand carried aloft by a strong wind. The sand particles are mostly confined to the lowest ten feet, and rarely rise more than fifty feet above the ground.

e. **Duststorm.** A severe weather condition characterized by strong winds and dust-filled air over an extensive area.

8.4 Present Weather Observing Standards

8.4.1 Qualifiers. Present weather qualifiers fall into two categories: intensity or proximity and descriptors. Qualifiers may be used in various combinations to describe weather phenomena.

a. **Intensity/Proximity.** The intensity qualifiers are: light, moderate, and heavy. The proximity qualifier is vicinity.

(1) **Intensity of Precipitation.** When more than one form of precipitation is occurring at a time or precipitation is occurring with an obscuration, the intensities determined shall be no greater than that which would be determined if any forms were occurring alone.

The intensity of precipitation shall be identified as light, moderate, or heavy in accordance with one of the following:

(a) **Intensity of Rain or Ice Pellets**. The intensity of rain and ice pellets shall be based on the criteria given in Table 8-1, Table 8-2, and Table 8-3.

Table 8-1. Intensity of Rain or Ice Pellets Based on Rate-of-Fall

Intensity	Criteria
Light	Up to 0.10 inch per hour; maximum 0.01 inch in 6 minutes.
Moderate	0.11 inch to 0.30 inch per hour; more than 0.01 inch to 0.03 inch in 6 minutes.
Heavy	More than 0.30 inch per hour; more than 0.03 inch in 6 minutes.

Table 8-2. Estimating Intensity of Rain

Intensity	Criteria
Light	From scattered drops that, regardless of duration, do not completely wet an exposed surface up to a condition where individual drops are easily seen.
Moderate	Individual drops are not clearly identifiable; spray is observable just above pavements and other hard surfaces.
Heavy	Rain seemingly falls in sheets; individual drops are not identifiable; heavy spray to height of several inches is observed over hard surfaces.

Table 8-3. Estimating Intensity of Ice Pellets

Intensity	Criteria
Light	Scattered pellets that do not completely cover an exposed surface regardless of duration. Visibility is not affected.
Moderate	Slow accumulation on ground. Visibility reduced by ice pellets to less than 7 statute miles.
Heavy	Rapid accumulation on ground. Visibility reduced by ice pellets to less than 3 statute miles.

(b) **Intensity of Snow and Drizzle.** The intensity of snow and drizzle shall be based on the reported surface visibility in accordance with Table 8-4 when occurring alone.

Table 8-4. Intensity of Snow or Drizzle Based on Visibility

Intensity	Criteria
Light	Visibility > 1/2 mile.
Moderate	Visibility > 1/4 mile but ≤ 1/2 mile.
Heavy	Visibility ≤ 1/4 mile.

(2) **Proximity.** Unless otherwise directed elsewhere in this Handbook, weather phenomena occurring beyond the point of observation (between 5 and 10 statute miles) shall be reported as (in the) vicinity.

b. **Descriptors.** Descriptors are qualifiers which further amplify weather phenomena and are used with certain types of precipitation and obscurations. The descriptor qualifiers are: shallow, partial, patches, low drifting, blowing, shower(s), thunderstorm, and freezing.

(1) **Shallow.** The descriptor shallow shall only be used to further describe fog that has little vertical extent (less than 6 feet).

(2) **Partial and Patches.** The descriptors partial and patches shall only be used to further describe fog that has little vertical extent (normally greater than or equal to 6 feet but less than 20 feet), and reduces horizontal visibility, but to a lesser extent vertically. The stars may often be seen by night and the sun by day.

(3) **Low Drifting.** When dust, sand, or snow is raised by the wind to less than 6 feet, "low drifting" shall be used to further describe the weather phenomenon.

(4) **Blowing.** When dust, sand, snow, and/or spray is raised by the wind to a height of 6 feet or more, "blowing" shall be used to further describe the weather phenomenon.

(5) **Shower(s).** Precipitation characterized by the suddenness with which they start and stop, by the rapid changes of intensity, and usually by rapid changes in the appearance of the sky.

(6) **Thunderstorm.** A local storm produced by a cumulonimbus cloud that is accompanied by lightning and/or thunder.

(7) **Freezing.** When fog is occurring and the temperature is below 0°C, "freezing" shall be used to further describe the phenomena. When drizzle and/or rain freezes upon impact and forms a glaze on the ground or other exposed objects, "freezing" shall be used to further describe the precipitation.

8.4.2 Weather Phenomena. Weather phenomena fall into three categories: precipitation, obscurations, and other phenomena. The three categories of weather phenomena shall be combined with the qualifiers listed in the preceding paragraphs, to identify present weather that is occurring.

8.5 Present Weather Reporting Standards

Present weather is reported when it is occurring at, or in the vicinity of, the station and at the time of observation. Unless directed elsewhere in the Handbook, the location of weather phenomena shall be reported as:

♦ "occurring at the station" when within 5 statute miles of the point(s) of observation.

♦ "in the vicinity of the station" when between 5 and 10 statute miles of the point(s) of observation.

♦ "distant from the station" when beyond 10 statute miles of the point(s) of observation.

With the exception of volcanic ash, low drifting dust, low drifting sand, low drifting snow, shallow fog, partial fog, and patches (of) fog, obscurations are reported only when the prevailing visibility is less than 7 statute miles or considered operationally significant. Volcanic ash shall always be reported when observed.

When more than one type of present weather are reported at the same time, present weather shall be reported in the following order:

♦ Tornadic activity--Tornado, Funnel Cloud, or Waterspout.
♦ Thunderstorm(s) with or without associated precipitation.
♦ Present weather in order of decreasing dominance, i.e., the most dominant type is reported first.
♦ Left-to-right in Table 8-5 (Columns 1 through 5).

The reporting notations given in Table 8-5 shall be used to report present weather. (For definitions of present weather, refer to Appendix A - Glossary).

8.5.1   Precipitation.   Precipitation shall be reported when occurring at the point of observation. Precipitation not occurring at the point of observation but within 10 statute miles shall be reported as showers in the vicinity.

   a. **Liquid Precipitation**

   (1)   **Drizzle.** (see paragraphs 12.6.8.a(1), 12.6.8.c(1), and 12.7.1.k).

   (2)   **Rain.** (see paragraphs 12.6.8.a(1), 12.6.8.c(1), and 12.7.1.k).

   (3)   **Rainshower(s).** (see paragraphs 12.6.8.a(1), 12.6.8.b(3), 12.6.8.c(1), and 12.7.1.k).

   b. **Freezing Precipitation**

   (1)   **Freezing Rain.** (see paragraphs 12.6.8.a(1), 12.6.8.b(5), 12.6.8.c(1), 12.7.1.k).

   (2)   **Freezing Drizzle.** (see paragraphs 12.6.8.a(1), 12.6.8.b(5), 12.6.8.c(1), and 12.7.1.k).

   c. **Solid Precipitation**

   (1)   **Snow.** (see paragraphs 12.6.8.a(1), 12.6.8.c(1), 12.7.1.k).

   (2)   **Snowshower(s).** (see paragraphs 12.6.8.a(1), 12.6.8.b(3), 12.6.8.c(1), and 12.7.1.k).

   (3)   **Blowing Snow.** (see paragraphs 12.6.8.a(1), 12.6.8.b(2), and 12.6.8.c(1)).

   (4)   **Low Drifting Snow.** (see paragraphs 12.6.8.a(1), 12.6.8.b(2), and 12.6.8.c(1)).

   (5)   **Snow Grains.** (see paragraphs 12.6.8.a(1), 12.6.8.c(1), 12.7.1.k).

   (6)   **Ice Crystals.** (see paragraphs 12.6.8.a(1), 12.6.8.c(1), and 12.7.1.k).

   (7)   **Ice Pellets.** (see paragraphs 12.6.8.c(1) and 12.7.1.k).

   Ice Pellet shower(s). (see paragraphs 12.6.8.b(3), 12.6.8.c(1), and 12.7.1.k).

   (8)   **Hail.** Hail shall be reported, at designated stations. (see paragraph 12.7.1.k and 12.7.1.n).

   Hail shower(s). (see paragraphs 12.6.8.b(3), 12.6.8.c(1), 12.7.1.k, and 12.7.1.n).

   (9)   **Small Hail** and/or **Snow Pellets.** (see paragraphs 12.6.8.c(1)).

   Small Hail and/or **Snow Pellets Shower(s).** (see paragraphs 12.6.8.b(3), 12.6.8.c(1), and 12.7.1.n).

   d. **Unknown Precipitation.** Unknown precipitation shall only be reported by automated stations to indicate precipitation of unknown type when the automated system cannot identify the precipitation with any greater precision (see paragraph 12.6.8.c(2)).

Table 8-5. Notations for Reporting Present Weather[1]

QUALIFIER		WEATHER PHENOMENA		
INTENSITY OR PROXIMITY 1	DESCRIPTOR 2	PRECIPITATION 3	OBSCURATION 4	OTHER 5
- Light Moderate[2] + Heavy VC In the Vicinity[3]	MI Shallow PR Partial BC Patches DR Low Drifting BL Blowing SH Shower(s) TS Thunderstorm FZ Freezing	DZ Drizzle RA Rain SN Snow SG Snow Grains IC Ice Crystals PL Ice Pellets GR Hail GS Small Hail and/or Snow Pellets UP Unknown Precipitation	BR Mist FG Fog FU Smoke VA Volcanic Ash DU Widespread Dust SA Sand HZ Haze PY Spray	PO Well-Developed Dust/Sand Whirls SQ Squalls FC Funnel Cloud Tornado Waterspout[4] SS Sandstorm DS Duststorm

1. The weather groups shall be constructed by considering columns 1 to 5 in the table above in sequence, i.e., intensity, followed by description, followed by weather phenomena, e.g., heavy rain shower(s) is coded as +SHRA
2. To denote moderate intensity no entry or symbol is used.
3. See paragraph 8.4.1.a.(2), 8.5, and 8.5.1 for vicinity definitions.
4. Tornadoes and waterspouts shall be coded as +FC.

8.5.2 Obscuration.

a. **Mist.** (see paragraph 12.6.8.d(1)).

b. **Fog.** (see paragraphs 12.6.8.a(2) and 12.6.8.d(1)).

  (1) **Shallow (Ground) Fog.** (see paragraphs 12.6.8.b(1) and 12.6.8.d(1)).

  (2) **Partial Fog.** (see paragraphs 12.6.8.b(1) and 12.6.8.d(2)).

  (3) **Patches (of) Fog.** (see paragraphs 12.6.8.b(1) and 12.6.8.d(2)).

  (4) **Freezing Fog.** (see paragraph 12.6.8.b(5) and 12.6.8.d(1)).

c. **Smoke.** (see paragraph 12.6.8.d(1)).

d. **Volcanic Ash.** (see paragraph 12.6.8.d(1)).

e. **Widespread Dust.** (see paragraph 12.6.8.d(1)).

  (1) **Blowing Dust.** (see paragraphs 12.6.8.a(1), 12.6.8.a(2), 12.6.8.b(2), and 12.6.8.d(1)).

  (2) **Low Drifting Dust.** (see paragraphs 12.6.8.b(2) and 12.6.8.d(1)).

f. **Sand.** (see paragraph 12.6.8.d(1)).

    (1) **Blowing Sand.** (see paragraphs 12.6.8.a(1), 12.6.8.a(2), 12.6.8.b(2), and 12.6.8.d(1)).

    (2) **Low Drifting Sand.** (see paragraphs 12.6.8.b(2) and 12.6.8.d(1)).

g. **Haze.** (see paragraph 12.6.8.d(1)).

h. **Blowing Spray.** (see paragraphs 12.6.8.b(2) and 12.6.8.d(3)).

8.5.3   **Other Weather Phenomena**

a. **Well-Developed Dust/Sand Whirls.** (see paragraphs 12.6.8.a(2) and 12.6.8.e(1)).

b. **Squalls.** (see paragraph 12.6.8.e(1)).

c. **Tornado, Waterspout, or Funnel Cloud.** (see paragraphs 12.6.8.a(1), 12.6.8.e(2), and 12.7.1.b).

d. **Sandstorm.** (see paragraphs 12.6.8.a(1) and 12.6.8.e(1)).

e. **Duststorm.** (see paragraphs (12.6.8.a(1) and 12.6.8.e(1)).

8.5.4   **Thunderstorm.** A thunderstorm occurring with or without accompanying precipitation shall be reported when observed to begin, to be in progress, or to end. In addition to reporting a thunderstorm in the body of the METAR/SPECI, remarks may be added to report the time, location, and movement of the storm (see paragraphs 8.5.5.c, 12.7.1.l, and 12.7.1.m).

a. **Beginning of Thunderstorm.** The beginning of a thunderstorm shall be reported as the earliest time:

    (1) thunder is heard;

    (2) lightning is observed at the station when the local noise level is sufficient to prevent hearing thunder; or

    (3) lightning is detected by an automated sensor.

b. **Ending of Thunderstorm.** The ending of a thunderstorm shall be reported as 15 minutes after the last occurrence of any of the above criteria.

8.5.5   **Beginning/Ending Times of Precipitation, Tornadic Activity, and Thunderstorms.**

a. **Precipitation.** At designated stations, the time precipitation begins or ends shall be reported to the nearest minute. The beginning and ending times shall be reported in the next METAR after the event. Beginning and ending times for separate periods shall be reported only if the intervening time exceeds 15 minutes (see paragraph 12.7.1.k).

b. **Tornadic Activity.** At designated stations, the time tornadic activity begins or ends shall be reported to the nearest minute. The beginning and ending times shall be reported in a SPECI and the next METAR after the event (see paragraphs 12.6.8.e(2) and 12.7.1.b).

c. **Thunderstorm.** At designated stations, the time thunderstorm(s) begins or ends shall be reported to the nearest minute. The beginning and ending times shall be reported in a SPECI and the next METAR after the event. Beginning and ending times of separate thunderstorm(s) shall be reported only in a METAR if the intervening time exceeds 15 minutes (see paragraphs 12.6.8.a(1), 12.7.1.l, and 12.7.1.m).

8.5.6 **Other Significant Weather Phenomena.** Observers shall be alert to weather phenomena that are visible from the station but not occurring at the station. Examples of such phenomena are fog banks, localized rain, snow blowing over runways, etc. These phenomena shall be reported whenever they are considered to be operationally significant. Volcanic eruptions shall be reported in remarks (see paragraph 12.7.1.a).

8.6 **Summary of Weather.** Table 8-6 contains a summary of the present weather observing and reporting standards according to the type of station.

Table 8-6. Summary of Present Weather Observing and Reporting Standards

Present Weather	Type of Station	
	Automated	Manual
Funnel Cloud (Tornadic Activity)	Augmented at designated stations.	Report FC, or +FC, and in remarks TORNADO, FUNNEL CLOUD, WATERSPOUT, time of beginning and time of ending, source, location, and direction of movement.
Thunderstorms	Augmented at designated stations.	Report TS, time of beginning/ending, location, and movement.
Hail	Augmented at designated stations	Report GR, time of beginning and time of ending, estimated size of largest hailstone in inches preceded by "GR".
Small hail and/or snow pellets	Augmented at designated stations.	Report GS, time of beginning and time of ending.
Obscurations	BR, FG and HZ may be reported.	Report BR, FG, PRFG, FU, DU, HZ, SA, BLSN, BLSA, BLDU, SS, DS, BLPY, and VA.
	N/A	Reports non-uniform weather and obscurations.
Precipitation	DZ, RA, SN, and UP may be reported.	Report RA, SHRA, DZ, FZRA, FZDZ, SN, SHSN, SG, GS, IC, GR, PL, and SHPL.
	May be reported as FZ.	Reports descriptor with precipitation.
	May report the intensity of precipitation as light, moderate, or heavy.	Reports the intensity of precipitation, other than IC, GR, and GS as light, moderate, or heavy.
	May report hourly accumulation of liquid precipitation.	May report hourly accumulation of liquid precipitation.
	May report 3-, 6-, and 24-hour accumulation of precipitation (water equivalent of solid).	May report 3-, 6-, and 24-hour accumulation of precipitation (water equivalent of solid).
	N/A	May report depth and accumulation of solid precipitation.
	N/A	Reports size of GR.
Squall	Report SQ.	Report SQ.

CHAPTER 9

SKY CONDITION

9.1 General

Sky condition is a description of the appearance of the sky. Sky condition may be evaluated either automatically by instrument or manually with or without instruments.

9.2 Scope

This chapter prescribes the standards for observing and reporting sky condition.

9.3 Sky Condition Parameters

Sky condition parameters are:

    a. **Sky cover.** The amount of the celestial dome hidden by clouds and/or obscurations.

    b. **Summation layer amount.** A categorization of the amount of sky cover at and below each reported layer.

    c. **Layer height.** The height of the bases of each reported layer of clouds and/or obscurations; or the vertical visibility into an indefinite ceiling.

    d. **Ceiling.** The lowest layer aloft reported as broken or overcast; or the vertical visibility into an indefinite ceiling.

    e. **Type of clouds.** The variety of clouds present.

9.4 Sky Condition Standards

9.4.1 Sky Condition Observing Standards. Sky condition shall be evaluated at all stations with this capability. Automated stations shall have the capability to evaluate sky condition from the surface to at least 12,000 feet. Observers at manual stations shall evaluate all clouds and obscurations visible; the 12,000 foot restriction shall not apply.

    a. **Layer Opacity.** All cloud layers and obscurations shall be considered as opaque.

    b. **Surface.** The surface shall be the assigned field elevation of the station. At stations where the field elevation has not been established, the surface shall be the ground elevation at the observation site.

    c. **Sky Cover.** Sky cover shall include any clouds or obscurations detected from the observing location.

    d. **Stratification of Sky Cover.** Sky cover shall be separated into layers with each layer containing clouds and/or obscurations (i.e., smoke, haze, fog, etc.) with bases at about the same height.

    e. **Evaluation of Interconnected Layers.** Clouds formed by the horizontal extension of swelling cumulus or cumulonimbus, that are attached to a parent cloud, shall be regarded as a separate layer only if their bases appear horizontal and at a different level from the parent cloud. Otherwise, the entire cloud system shall be regarded as a single layer at a height corresponding to the base of the parent cloud.

f. **Sky Condition Range, Accuracy, and Resolution.** The required range, accuracy, and resolution for sky condition is listed in Appendix C.

9.4.2 Sky Cover

a. **Clear Skies.** When no clouds or obscurations are observed or detected from the point of observation.

b. **Layer Amounts.** The amount of sky cover for each layer shall be the eighths (or oktas) of sky cover attributable to clouds or obscurations (i.e., smoke, haze, fog, etc.) in the layer being evaluated.

c. **Summation Layer Amount.** The sky cover summation amount for any given layer is the sum of the sky cover for the layer being evaluated plus the sky cover of all lower layers including obscurations. Portions of layers aloft detected through lower layers aloft shall not increase the summation amount of the higher layer. No layer can have a summation amount greater than 8/8ths.

d. **Variable Amounts of Sky Cover.** The sky cover shall be considered variable if it varies by one or more reportable values (FEW, SCT, BKN, or OVC) during the period it is being evaluated.

9.4.3 Obscuration. The portion of sky (including higher clouds, the moon, or stars) hidden by weather phenomena either surface-based or aloft.

9.4.4 Vertical Visibility. Vertical visibility shall be either:

a. The distance that an observer can see vertically into an indefinite ceiling;

b. The height corresponding to the top of a ceiling light projector beam;

c. The height at which a ceiling balloon completely disappears during the presence of an indefinite ceiling; or

d. The height determined by the sensor algorithm at automated stations.

9.4.5 Ceiling. The ceiling shall be the lowest layer aloft reported as broken or overcast. If the sky is totally obscured, the vertical visibility shall be the ceiling.

9.4.6 Significant Clouds and Cloud Types. Cloud types shall be identified in accordance with the WMO International Cloud Atlas-Volumes I and II, the WMO *Abridged International Cloud Atlas*, or agency observing aids for cloud identification. Cumulonimbus, including cumulonimbus mammatus, towering cumulus, altocumulus castellanus, standing lenticular, or rotor clouds are significant clouds.

9.4.7 Height of Sky Cover. A ceilometer, if available, or ceiling light, or known heights of unobscured portions of abrupt, isolated objects within 1 1/2 statute miles of a runway shall be used to measure the height of layers aloft. Otherwise, an alternative method shall be used to estimate the heights. The height may be estimated by using a ceiling balloon, pilot report, other agency guidelines, or observer experience.

a. **Indefinite Ceiling Height (Vertical Visibility).** The height into an indefinite ceiling shall be the vertical visibility measured in hundreds of feet.

b. **Height of Layers.** The height of a layer shall be the height of the cloud bases or obscurations for the layer being evaluated. Layers of clouds that are 50 feet or less above the surface shall be observed as layers with a height of zero. When the height of a ceiling layer increases and decreases rapidly by the amounts given in Table 9-2, during the period of evaluation, it shall be considered variable and the ascribed height shall be the average of all the varying values. At mountain stations, clouds below the level of the station may be observed.

Table 9-1. Criteria for Variable Ceiling

Ceiling (feet)	Variation (feet)
≤ 1,000	≥200
>1,000 and ≤2,000	≥400
>2,000 and <3,000	≥500

9.5 Sky Cover Reporting Standards

9.5.1 Frequency for Sky Cover. Sky cover shall be included in all reports.

9.5.2 Layer Amount.

The amount of sky cover reported for each layer shall be based on the summation layer amount for that layer. The amount shall be reported using the reportable contractions given in Table 9-2.

Automated stations shall report no more than three layers. The selection of layers reported shall be made in accordance with Table 9-3. Manual stations shall report no more than six layers. If more than six layers are observed, then use Table 9-3 to determine which layers are to be reported. Additionally, all layers with associated cumulonimbus or towering cumulus shall be identified by appending the contractions **CB** and **TCU**, respectively.

Sky condition shall be reported in an ascending order up to the first overcast layer. Layers above 12,000 feet are not reported by automated sky condition sensors. At mountain stations, if the cloud layer is below station level, the height of the layer shall be reported as ///.

Table 9-2. Reportable Contractions for Sky Cover

Reportable Contraction	Meaning	Summation Amount of Layer
VV	Vertical Visibility	8/8
SKC or CLR[1]	Clear	0
FEW[2]	Few	1/8 - 2/8
SCT	Scattered	3/8 - 4/8
BKN	Broken	5/8 - 7/8
OVC	Overcast	8/8

1. The abbreviation **CLR** shall be used at automated stations when no layers at or below 12,000 feet are reported; the abbreviation **SKC** shall be used at manual stations when no layers are reported.
2. Any layer amount less than 1/8 is reported as FEW.

Table 9-3. Priority for Reporting Layers

Priority	Layer Description
1	lowest few layer.
2	lowest broken layer.
3	overcast layer.
4	lowest scattered layer.
5	second lowest scattered layer.
6	second lowest broken layer.
7	highest broken layer.
8	highest scattered layer.

9.5.3   Units of Measure for Heights. Heights of sky cover shall be evaluated in feet above the surface.

9.5.4   Reportable Values for Sky Cover Height. The reportable values of sky cover height are hundreds of feet. The reportable value increments are given in Table 9-4.

Table 9-4. Increments of Reportable Values of Sky Cover Height

Range of Height Values (feet)	Reportable Increment (feet)
≤5,000	To nearest 100
>5,000 but ≤10,000	To nearest 500
>10,000	To nearest 1,000

9.5.5   Layer Heights. Heights of layers shall be reported in hundreds of feet, rounded to the nearest reportable increment. When a value falls halfway between two reportable increments, the lower value shall be reported. When a cloud layer is 50 feet or less above the surface, the height shall be reported as 000 (see paragraph 9.4.7.b).

9.5.6   Obscuration. When a portion of the celestial dome is obscured, the obscuration (amount of sky cover hidden by the weather phenomena) shall be reported (see paragraph 9.4.3 and Table 9-1). The obscuration shall also be reported as a remark (see paragraph 12.7.1.q).

9.5.7   Variable Ceiling. When the height of the ceiling layer is variable, and the ceiling layer is below 3,000 feet, a remark shall be included in the report giving the range of variability (see paragraphs 9.4.7.b and 12.7.1.p).

9.5.8   Ceiling Height at a Second Location. When automated stations use meteorological discontinuity ceilometer(s), remarks shall be added to identify ceiling height conditions at the second location which differ from the ceiling height in the body of the report (see paragraph 12.7.1.t).

9.5.9   Variable Sky Condition. Variable sky conditions shall be indicated in the remarks of the report (see paragraph 12.7.1.r).

9.5.10 Significant Cloud Types. Significant cloud types shall be indicated in the remarks of the report (see paragraph 12.7.1.s).

9.6 Summary of Sky Condition Observing and Reporting Standards

Table 9-5 summarizes the sky condition observing and reporting at each category of station.

Table 9-5. Summary of Sky Condition Observing and Reporting Standards

Parameter	Reporting Standard
Sky Cover (General)	Sky condition shall be included in all reports.
Height/Number of layers	Report a maximum of three layers at automated stations; otherwise, a maximum of six layers at manual stations.
Variable sky condition	Not evaluated at automated stations.
Variable ceiling height	Evaluated at all stations.
Ceiling height at a second location	Evaluated at automated stations with multiple sensors.
Cloud Types	Not evaluated at automated stations.

CHAPTER 10

TEMPERATURE AND DEW POINT

10.1 General

Temperature is a measure of hotness or coldness. On a daily basis, temperature is one of the most widely monitored and disseminated weather parameters obtained from the surface observation.

10.2 Scope

This chapter prescribes the standards for observing and reporting temperature and dew point. The chapter also defines maximum and minimum temperature and prescribes appropriate standards.

10.3 Temperature and Dew Point Parameters

a. **Temperature.** The degree of hotness or coldness of the ambient air as measured by any suitable instrument.

b. **Dew point.** The temperature to which a given parcel of air must be cooled at constant pressure and constant water-vapor content in order for saturation to occur.

c. **Maximum temperature.** The highest temperature recorded/measured during a specified time period.

d. **Minimum temperature.** The lowest temperature recorded/measured during a specified time period.

10.4 Temperature and Dew Point Observing Standards

The method of obtaining temperature and dew point varies according to the system in use at the station. The data may be read directly from digital or analog readouts, or calculated from other measured values.

10.4.1 Temperature and Dew Point Sensor Range. The range for the temperature and dew point sensors shall be determined by the responsible agency considering the local climatology (Appendix C).

10.4.2 Temperature. Temperature shall be determined to the nearest tenth of a degree Celsius at all stations.

10.4.3 Dew Point. At designated stations, dew point shall be determined to the nearest tenth of a degree Celsius with respect to water at all temperatures.

10.4.4 Maximum and Minimum Temperature. At designated stations, maximum and minimum temperatures that occurred in the previous 6 hours shall be determined to the nearest tenth of a degree Celsius for the 0000, 0600, 1200, and 1800 UTC observations. The maximum and minimum temperatures for the previous 24 hours shall be determined to the nearest tenth of a degree Celsius for the 0000 LST observation.

10.5 Temperature and Dew Point Reporting Standards

10.5.1 Resolution for Temperature and Dew Point. The reporting resolution for the temperature and the dew point in the body of the report shall be whole degrees Celsius. The reporting resolution for the temperature and dew point in the remarks section of the report shall be to the nearest tenth of a degree Celsius. Dew point shall be calculated with respect to water at all temperatures.

10.5.2   Maximum and Minimum Temperatures. At designated stations, maximum and minimum temperatures shall be reported as additive data in the 0000, 0600, 1200, and 1800 UTC and 0000 LST observations (see paragraphs 12.7.2.e, 12.7.2.f, and 12.7.2.g).

10.5.3   Reporting Procedures. Temperature and dew point are reported in the body of the report in accordance with paragraph 12.6.10. Temperature and dew point in the remarks section shall only be reported in METARs (see paragraph 12.7.2.d). Maximum and minimum temperatures shall be reported in the remarks section of the METAR in accordance with paragraphs 12.7.2.e, 12.7.2.f, and 12.7.2.g.

10.6   Summary of Temperature and Dew Point Observing and Reporting Standards

Table 10-1 summarizes the temperature and dew point observing and reporting procedures.

Table 10-1. Summary of Temperature and Dew Point Observing and Reporting Standards

Parameter	Section of Report	
	Body of METAR & SPECI	Remarks of METAR
Temperature	Reported in whole degrees Celsius at all stations.	Reported to tenths of degrees Celsius at designated stations.
Dew Point	Reported in whole degrees Celsius at designated stations.	Reported to tenths of degrees Celsius at designated stations.
Maximum and Minimum Temperatures		Designated stations report at 0000, 0600, 1200, and 1800 UTC.
24-hour Maximum and Minimum Temperatures		Designated stations report at 0000 LST.

CHAPTER 11

PRESSURE

11.1 General

Atmospheric pressure is the force exerted by the atmosphere at a given point. In this chapter, the term "barometric pressure" refers to the actual pressure sensor value. The sensor value may be an altimeter setting, station pressure, or simply a direct pressure value without applied corrections depending on the type of sensor.

11.2 Scope

This chapter prescribes the standards for observing and reporting atmospheric pressure data.

11.3 Pressure Parameters

a. **Station pressure.** The atmospheric pressure at the designated station elevation.

b. **Altimeter setting.** The pressure value to which an aircraft altimeter scale is set so that it will indicate the altitude above mean sea level of an aircraft on the ground at the location for which the value was determined.

c. **Sea-level pressure.** A pressure value obtained by the theoretical reduction of barometric pressure to sea level. Where the Earth's surface is above sea level, it is assumed that the atmosphere extends to sea level below the station and that the properties of that hypothetical atmosphere are related to conditions observed at the station.

11.4 Pressure Observing Standards

11.4.1 Barometer Comparisons. Each agency shall establish an agency standard barometer traceable to the standard of the National Institute of Standards and Technology. Each agency shall also establish a system of routine barometer comparisons to determine corrections required to keep the station's pressure sensors within the required accuracy (see Appendix C).

11.4.2 Atmospheric Pressure. The various pressure parameters shall be determined from the barometric pressure after appropriate corrections are applied. The method used shall depend on the type of sensor and the available computational aids. These aids may be systems that result in a direct readout of the desired parameter, pressure reduction calculators, or tables. Designated stations may use constants to convert measured pressure to the desired pressure parameter.

11.4.3 Station Pressure. Station pressure shall be determined by adjusting the corrected barometric pressure to compensate for the difference between the height of the barometer and the designated station elevation.

11.4.4 Sea-Level Pressure. At designated stations, sea-level pressure shall be computed by adjusting the station pressure to compensate for the difference between the station elevation and sea-level. This adjustment shall be based on the station elevation and the 12-hour mean temperature at the station. The 12-hour mean temperature shall be the average of the present ambient temperature and the ambient temperature 12 hours ago.

Stations within ± 50 feet of sea-level may be authorized by their agency to use a constant value to adjust station pressure to sea-level pressure. Otherwise, stations shall use reduction ratios provided by their responsible agency to calculate sea-level pressure.

11.4.5   Altimeter Setting. The altimeter setting shall be determined either directly from an altimeter setting indicator or computed from the station pressure by applying a correction for the difference between the station elevation and field elevation in the standard atmosphere. Where this difference is 30 feet or less, agencies may authorize the use of a constant correction.

11.4.6   Pressure Change (Rising/Falling). At designated stations, the pressure calculated for each report shall be examined to determine if a pressure change is occurring. If the pressure is rising or falling at a rate of at least 0.06 inch per hour and the pressure change totals 0.02 inch or more at the time of the observation, a pressure change remark shall be reported (see paragraph 12.7.1.u).

11.4.7   Pressure Tendency. Designated stations shall include pressure tendency data in each 3- and 6-hourly report. The pressure tendency includes two parts: the characteristic (an indication of how the pressure has been changing over the past three hours) and the amount of the pressure change in the past three hours. The characteristic shall be based on the observed or recorded (barogram trace) changes in pressure over the past three hours. The amount of pressure change is the absolute value of the change in station pressure or altimeter setting in the past three hours converted to tenths of hectopascals.

11.5   Pressure Reporting Standards

11.5.1   Rounding Pressure Values. When computations of pressure values require that a number be rounded to comply with standards on reportable values, the number shall be rounded down to the next reportable value. For example, an altimeter reading of 29.248 inches becomes 29.24 and a station pressure reading of 29.249 inches becomes 29.245.

11.5.2   Units of Measure. Table 11-1 lists the units of measure for pressure parameters.

Table 11-1. Units of Measure of Pressure Parameters

Parameter	Units of Measure
Altimeter Setting	Inches of Mercury
Sea-Level Pressure	Hectopascals
Station Pressure	Inches of Mercury

11.5.3   Altimeter Setting. Altimeter setting shall be reported in all reports (see paragraph 12.6.11).

11.5.4   Sea-Level Pressure. At designated stations, sea-level pressure shall be included in the remarks section of all METARs (see paragraph 12.7.1.v).

11.5.5   Remarks. At designated stations, the pressure change remarks (PRESRR or PRESFR) shall be reported if occurring at the time of observation (see paragraph 12.7.1.u). The pressure tendency group shall only be included in 3- and 6-hourly reports (see paragraph 12.7.2.h).

11.6 Summary of Pressure Observing and Reporting Standards

Table 11-2 summarizes the pressure observing and reporting standards.

Table 11-2. Summary of Pressure Observing and Reporting Standards

Parameter	Reporting Standard
Altimeter Setting	Reported in inches of mercury at all stations.
Sea-level pressure	Reported in hectopascals at designated stations.
Remarks: Rising Rapidly Falling Rapidly	Reported at designated stations.
Pressure Tendency	Reported at designated stations.

CHAPTER 12

CODING

12.1 General

The coding standards in this chapter conform to WMO Code Forms METAR FM 15-IX Ext. and SPECI FM 16-IX Ext. and the United States Exceptions filed with the WMO. These exceptions reflect national observing practices which differ from practices outlined in the WMO *Manual on Codes* No. 306.

12.2 Scope

This chapter contains the standards for coding a weather observation in either aviation routine weather report (METAR) and/or aviation selected special weather report (SPECI) format.

12.3 METAR/SPECI Code

> **METAR** or **SPECI**_**CCCC**_**YYGGggZ**_**AUTO** or **COR**_**dddff(f)Gf$_m$f$_m$(f$_m$)KT**_**d$_n$d$_n$d$_n$Vd$_x$d$_x$d$_x$**_
> **VVVVVSM**_**[RD$_R$D$_R$/V$_R$V$_R$V$_R$V$_R$FT** or **RD$_R$D$_R$/V$_N$V$_N$V$_N$V$_N$VV$_X$V$_X$V$_X$V$_X$FT]**_**w'w'**_**[N$_s$N$_s$N$_s$h$_s$h$_s$h$_s$** or **VVh$_s$h$_s$h$_s$**
> or **SKC/CLR]**_**T'T'/T'$_d$T'$_d$**_**AP$_H$P$_H$P$_H$P$_H$**_**RMK**_(**Automated, Manual, Plain Language**)_
> (**Additive Data and Automated Maintenance Indicators**)

METAR/SPECI has two major sections: the Body (consisting of a maximum of 11 groups) and the Remarks (consisting of 2 categories). Together, the body and remarks make up the complete METAR/SPECI. In general, the remarks are coded in the order depicted above and established in the remainder of this chapter.

12.4 Format and Content of the METAR/SPECI

a. Body of report.

(1) Type of Report - **METAR/SPECI**
(2) Station Identifier - **CCCC**
(3) Date and Time of Report - **YYGGggZ**
(4) Report Modifier - **AUTO/COR**
(5) Wind - **dddff(f)Gf$_m$f$_m$(f$_m$)KT**_**d$_n$d$_n$d$_n$Vd$_x$d$_x$d$_x$**
(6) Visibility - **VVVVVSM**
(7) Runway Visual Range - **RD$_R$D$_R$/V$_R$V$_R$V$_R$V$_R$FT** or **RD$_R$D$_R$/V$_N$V$_N$V$_N$V$_N$VV$_X$V$_X$V$_X$V$_X$FT**
(8) Present Weather - **w'w'**
(9) Sky Condition - **N$_s$N$_s$N$_s$h$_s$h$_s$h$_s$** or **VVh$_s$h$_s$h$_s$** or **SKC/CLR**
(10) Temperature and Dew Point - **T'T'/T'$_d$T'$_d$**
(11) Altimeter - **AP$_H$P$_H$P$_H$P$_H$**

b. Remarks section of report--RMK

(1) Automated, Manual, and Plain Language
(2) Additive and Maintenance Data

The underline character "_" indicates a required space between the groups. If a group is not reported, the preceding space is also not reported. In addition to the format given, agencies shall provide for the inclusion of any special Beginning-of-Message, End-of-Message, or End-of-Transmission signals required by their communications system.

The actual content of a METAR or SPECI depends on the observation program at the individual station. At designated stations, the 0000, 0600, 1200, and 1800 Coordinated Universal Time (UTC) METAR's include additional data specified by the responsible agency and are known as 6-hourly reports. At designated stations, the 0300, 0900, 1500, and 2100 UTC METAR's are known as 3-hourly reports and also contain additional information specified by the responsible agency.

12.5 Coding Missing Data in METAR/SPECI

When an element does not occur, or cannot be observed, the corresponding group and preceding space are omitted from that particular report.

12.6 Coding the Body of the METAR/SPECI

12.6.1 Type of Report (METAR and SPECI). The type, **METAR** or **SPECI**, shall be included in all reports. The type of report shall be separated from elements following it by a space. Whenever SPECI criteria are met at the time of the routine METAR, the type of report shall be METAR.

12.6.2 Station Identifier (CCCC). The station identifier, **CCCC**, shall be included in all reports to identify the station to which the coded report applies. The station identifier shall consist of four alphabetic-only characters if the METAR/SPECI is transmitted long-line. The agency with operational control when the station is first established shall be responsible for coordinating the location identifier with the FAA. A list of approved identifiers can be found in the FAA Manual 7350 Series, *Location Identifiers*.

12.6.3 Date and Time of Report (YYGGggZ). The date, **YY**, and time, **GGgg**, shall be included in all reports. The time shall be the actual time of the report or when the criteria for a SPECI is met or noted (see paragraph 2.6.4). If the report is a correction to a previously disseminated report, the time of the corrected report shall be the same time used in the report being corrected. The date and time group always ends with a **Z** indicating Zulu time (or UTC). For example, METAR KDCA 210855Z would be the 0900 scheduled report from station KDCA taken at 0855 UTC on the 21st of the month.

12.6.4 Report Modifier (AUTO or COR). The report modifier, **AUTO,** identifies the METAR/SPECI as a fully automated report with no human intervention or oversight. In the event of a corrected METAR or SPECI, the report modifier, **COR**, shall be substituted in place of **AUTO**.

12.6.5 Wind Group (dddff(f)G$f_m f_m (f_m)$KT_$d_n d_n d_n$V$d_x d_x d_x$). The standards for observing and reporting wind are described in Chapter 5.

The wind direction, **ddd**, shall be coded in tens of degrees using three figures. Directions less than 100 degrees shall be preceded with a "0". For example, a wind direction of 90° is coded as "090". The wind speed, **ff(f)**, shall be coded in two or three digits immediately following the wind direction. The wind speed shall be coded, in whole knots, using the units and tens digits and, if required, the hundreds digit. Speeds of less than 10 knots shall be coded using a leading zero. The wind group shall always end with **KT** to indicate that wind speeds are reported in knots. For example, a wind speed of 8 knots shall be coded "08KT"; a wind speed of 112 knots shall be coded "112KT".

    a. **Gust.** Wind gusts shall be coded in the format, **G$f_m f_m (f_m)$** (see paragraphs 5.4.4 and 5.5.4). The wind gust shall be coded in two or three digits immediately following the wind speed. The wind gust shall be coded, in whole knots, using the units and tens digits and, if required, the hundreds digit. For example, a wind from due west at 20 knots with gusts to 35 knots would be coded "27020G35KT".

    b. **Variable Wind Direction (Speeds 6 knots or less).** Variable wind direction with wind speed 6 knots or less may be coded as **VRB** in place of the **ddd** (see paragraphs 5.4.2 and 5.5.3). For example, if the wind is variable at three knots, it would be coded "VRB03KT".

    c. **Variable Wind Direction (Speeds greater than 6 knots).** Variable wind direction with wind speed greater than 6 knots shall be coded in the format, **$d_n d_n d_n$V$d_x d_x d_x$**. The variable wind direction group shall immediately follow the wind group (see paragraphs 5.4.2 and 5.5.3). The directional variability shall be coded in a clockwise direction. For example, if the wind is variable from 180° to 240° at 10 knots, it would be coded "21010KT 180V240".

    d. **Calm Wind.** Calm wind shall be coded as "00000KT" (see paragraph 5.5.2).

12.6.6 **Visibility Group (VVVVVSM).** The standards for observing and reporting visibility are described in Chapter 6.

The surface visibility, **VVVVVSM**, shall be coded in statute miles using the values listed in Table 12-1. A space shall be coded between whole numbers and fractions of reportable visibility values. The visibility group shall always end with **SM** to indicate that the visibility is in statute miles. For example, a visibility of one and a half statute miles would be coded "1 1/2SM".

Automated stations shall use an **M** to indicate "less than" when reporting visibility. For example, "M1/4SM" means a visibility of less than one-quarter statute mile.

Table 12-1. Reportable Visibility Values

Source of Visibility Report							
Automated			Manual				
M1/4	2	9[a]	0	5/8	1 5/8	4	12
1/4	2 1/2	10	1/16	3/4	1 3/4	5	13
1/2	3		1/8	7/8	1 7/8	6	14
3/4	4		3/16	1	2	7	15
1	5		1/4	1 1/8	2 1/4	8	20
1 1/4	6[a]		5/16	1 1/4	2 1/2	9	25
1 1/2	7		3/8	1 3/8	2 3/4	10	30
1 3/4	8[a]		1/2	1 1/2	3	11	35[b]

a. These values may not be reported by some automated stations.
b. Further values in increments of 5 statute miles may be reported, i.e., 40, 45, 50, etc

12.6.7. **Runway Visual Range Group ($RD_RD_R/V_RV_RV_RV_RFT$ or $RD_RD_R/V_nV_nV_nV_nVV_xV_xV_xV_xFT$).** The standards for observing and reporting Runway Visual Range (RVR) are described in Chapter 7.

a. RVR shall be coded in the format $RD_RD_R/V_RV_RV_RV_RFT$, where **R** indicates that the runway number follows, D_RD_R is the runway number (an additional D_R may be used for runway approach directions, such as **R** for right, **L** for left, and **C** for center), $V_RV_RV_RV_R$ is the constant reportable value, and **FT** indicates that units of measurement are feet. A solidus "/" without spaces separates the runway number from the constant reportable value. For example, an RVR value for runway 01L of 800 feet would be coded "R01L/0800FT".

b. RVR that is varying shall be coded in the format, $RD_RD_R/V_nV_nV_nV_nVV_xV_xV_xV_xFT$, where **R** indicates that the runway number follows, D_RD_R is the runway number (an additional D_R may be used for runway approach directions, such as **R** for right, **L** for left, and **C** for center), $V_nV_nV_nV_n$ is the lowest reportable value in feet, **V** separates lowest and highest visual range values, $V_xV_xV_xV_x$ is the highest reportable value, and **FT** indicates that units of measurement are feet. A solidus "/" without spaces separates the runway number from the reportable values. For example, the 10-minute RVR for runway 01L varying between 600 and 1,000 feet would be coded "R01L/0600V1000FT".

c. The values shall be based on light setting 5 at manual stations regardless of the light setting actually in use (see Appendix D). RVR values shall be coded in increments of 100 feet up to 1,000 feet, increments of 200 feet from 1,000 feet to 3,000 feet, and increments of 500 feet from 3,000 feet to 6,000 feet. Manual RVR shall not be reported below 600 feet. For automated stations, RVR may be reported from up to four designated runways.

d. If the RVR is less than its lowest reportable value, the $V_R V_R V_R V_R$ or $V_n V_n V_n V_n$ groups shall be preceded by **M**. If the RVR is greater than its highest reportable value, the $V_R V_R V_R V_R$ or $V_x V_x V_x V_x$ groups shall be preceded by a **P**. For example, an RVR for runway 01L of less than 600 feet will be coded "R01L/M0600FT"; an RVR for runway 27 of greater than 6,000 feet will be coded "R27/P6000FT".

12.6.8 **Present Weather Group (w'w').** The standards for observing and reporting present weather are described in Chapter 8.

The appropriate notations found in Table 12-2 shall be used to code present weather.

Table 12-2. Notations for Reporting Present Weather[1]

QUALIFIER		WEATHER PHENOMENA		
INTENSITY OR PROXIMITY 1	DESCRIPTOR 2	PRECIPITATION 3	OBSCURATION 4	OTHER 5
- Light Moderate[2] + Heavy **VC** In the Vicinity[3]	**MI** Shallow **PR** Partial **BC** Patches **DR** Low Drifting **BL** Blowing **SH** Shower(s) **TS** Thunderstorm **FZ** Freezing	**DZ** Drizzle **RA** Rain **SN** Snow **SG** Snow Grains **IC** Ice Crystals **PL** Ice Pellets **GR** Hail **GS** Small Hail and/or Snow Pellets **UP** Unknown Precipitation	**BR** Mist **FG** Fog **FU** Smoke **VA** Volcanic Ash **DU** Widespread Dust **SA** Sand **HZ** Haze **PY** Spray	**PO** Well-Developed Dust/Sand Whirls **SQ** Squalls **FC** Funnel Cloud Tornado Waterspout[4] **SS** Sandstorm **DS** Duststorm

1. The weather groups shall be constructed by considering columns 1 to 5 in the table above in sequence, i.e., intensity, followed by description, followed by weather phenomena, e.g., heavy rain shower(s) is coded as +SHRA
2. To denote moderate intensity no entry or symbol is used.
3. See paragraph 8.4.1.a.(2), 8.5, and 8.5.1 for vicinity definitions.
4. Tornadoes and waterspouts shall be coded as +FC.

The following general rules apply when coding present weather for a METAR or SPECI:

- ◆ Weather occurring at the point of observation (at the station) or in the vicinity of the station shall be coded in the body of the report; weather observed but not occurring at the point of observation (at the station) or in the vicinity of the station shall be coded in Remarks.

- ◆ With the exceptions of volcanic ash, low drifting dust, low drifting sand, low drifting snow, shallow fog, partial fog, and patches (of) fog, an obscuration shall be coded in the body of the report if the surface visibility is less than 7 miles or considered operationally significant. Volcanic ash shall always be coded when observed.

- ◆ Separate groups shall be used for each type of present weather. Each group shall be separated from the other by a space. METAR/SPECI shall contain no more than three present weather groups.

- ◆ The weather groups shall be constructed by considering columns 1 to 5 in Table 12-2 in sequence, i.e., intensity, followed by description, followed by weather phenomena, e.g., heavy rain shower(s) is coded as +SHRA.

a. **Intensity or Proximity Qualifier.**

(1) Intensity shall be coded with precipitation types, except ice crystals (**IC**), hail (**GR** or **GS**), and unknown precipitation (**UP**) including those associated with a thunderstorm (**TS**) and those of a showery nature (**SH**). Tornadoes and waterspouts shall be coded as +**FC**. No intensity shall be ascribed to the obscurations of blowing dust (**BLDU**), blowing sand (**BLSA**), and blowing snow (**BLSN**). Only moderate or heavy intensity shall be ascribed to sandstorm (**SS**) and duststorm (**DS**).

(2) The proximity qualifier for vicinity, **VC**, (weather phenomena observed in the vicinity of but not at the point(s) of observation) shall be coded in combination with thunderstorm (**TS**), fog (**FG**), shower(s) (**SH**), well-developed dust/sand whirls (**PO**), blowing dust (**BLDU**), blowing sand (**BLSA**), blowing snow (**BLSN**), sandstorm (**SS**), and duststorm (**DS**). Intensity qualifiers shall not be coded with **VC**.

VCFG shall be coded to report any type of fog in the vicinity of the point(s) of observation.

Precipitation not occurring at the point of observation but within 10 statute miles shall be coded as showers in the vicinity (**VCSH**).

b. **Descriptor Qualifier.** Only one descriptor shall be coded for each weather phenomena group, e.g., "-FZDZ". Mist (**BR**) shall not be coded with any descriptor.

(1) The descriptors shallow (**MI**), partial (**PR**), and patches (**BC**) shall only be coded with **FG**, e.g., "MIFG".

(2) The descriptors low drifting (**DR**) and blowing (**BL**) shall only be coded with dust (**DU**), sand (**SA**), and snow (**SN**), e.g., "BLSN" or "DRSN". **DR** shall be coded for **DU**, **SA**, or **SN** raised by the wind to less than six feet above the ground.

When blowing snow is observed with snow falling from clouds, both phenomena are reported, e.g., "SN BLSN". If there is blowing snow and the observer cannot determine whether or not snow is also falling, then **BLSN** shall be reported. **PY** shall be coded only with blowing (**BL**).

(3)   The descriptor shower(s) (**SH**) shall be coded only with one or more of the precipitation types of rain (**RA**), snow (**SN**), ice pellets (**PL**), small hail (**GS**), or large hail (**GR**). The **SH** descriptor indicates showery-type precipitation. When any type of precipitation is coded with **VC**, the intensity and type of precipitation shall not be coded.

(4)   The descriptor thunderstorm (**TS**) may be coded by itself, i.e., a thunderstorm without associated precipitation, or it may be coded with the precipitation types of rain (**RA**), snow (**SN**), ice pellets (**PL**), small hail and/or snow pellets (**GS**), or hail (**GR**). For example, a thunderstorm with snow and small hail and/or snow pellets would be coded as "TSSNGS". **TS** shall not be coded with **SH**.

(5)   The descriptor freezing (**FZ**) shall only be coded in combination with fog (**FG**), drizzle (**DZ**), or rain (**RA**), e.g., "FZRA". **FZ** shall not be coded with **SH**.

c.   **Precipitation.** Up to three types of precipitation may be coded in a single present weather group. They shall be coded in order of decreasing dominance based on intensity.

(1)   Drizzle shall be coded as **DZ**; rain shall be coded as **RA**; snow shall be coded as **SN**; snow grains shall be coded as **SG**; ice crystals shall be coded as **IC**; ice pellets shall be coded as **PL**, hail shall be coded as **GR**, and small hail and/or snow pellets shall be coded as **GS**.

(2)   At automated stations, precipitation of unknown type shall be coded as **UP**.

d.   **Obscuration.**

(1)   Mist shall be coded as **BR**; fog shall be coded as **FG**; smoke shall be coded as **FU**; volcanic ash shall be coded as **VA**; widespread dust shall be coded as **DU**; sand shall be coded as **SA**; and haze shall be coded as **HZ**.

(2)   Shallow fog (**MIFG**), patches (of) fog (**BCFG**), and partial fog (**PRFG**) may be coded with prevailing visibility of 7 statute miles or greater.

(3)   Spray (**PY**) shall be coded only as **BLPY**.

e.   **Other Weather Phenomena**

(1)   Well-developed dust/sand whirls shall be coded as **PO**; squalls shall be coded as **SQ**; sandstorm shall be coded as **SS**; and duststorm shall be coded as **DS**.

(2)   Tornadoes and waterspouts shall be coded as **+FC**. Funnel clouds shall be coded as **FC**.

12.6.9   <u>Sky Condition Group</u> ($N_sN_sN_sh_sh_sh_s$ or $VVh_sh_sh_s$ or **SKC/CLR**). The standards for observing and reporting sky condition are described in Chapter 9.

a.   Sky condition shall be coded in the format, $N_sN_sN_sh_sh_sh_s$, where $N_sN_sN_s$ is the amount of sky cover and $h_sh_sh_s$ is the height of the layer. There shall be no space between the amount of sky cover and the height of the layer. Sky condition shall be coded in an ascending order up to the first overcast layer. At mountain stations, if the layer is below station level, the height of the layer shall be coded as ///.

b.   Vertical visibility shall be coded in the format, $VVh_sh_sh_s$, where **VV** identifies an indefinite ceiling and $h_sh_sh_s$ is the vertical visibility into the indefinite ceiling (see paragraphs 9.4.4, 9.4.7, and 9.5.5). There shall be no space between the group identifier and the vertical visibility.

c.   Clear skies shall be coded in the format, **SKC** or **CLR**, where **SKC** is the abbreviation used by manual stations to indicate no layers are present and **CLR** is the abbreviation used by automated stations to indicate no layers are detected at or below 12,000 feet (see paragraph 9.5.4).

Each layer shall be separated from other layers by a space. The sky cover for each layer reported shall be coded by using the appropriate reportable contraction from Table 12-3. The report of clear skies (**SKC** or **CLR**) are complete layer reports within themselves. The abbreviations **FEW**, **SCT**, **BKN**, and **OVC** shall be followed, without a space, by the height of the layer.

Table 12-3. Contractions for Sky Cover

Reportable Contraction	Meaning	Summation Amount of Layer
VV	Vertical Visibility	8/8
SKC or CLR[1]	Clear	0
FEW[2]	Few	1/8 - 2/8
SCT	Scattered	3/8 - 4/8
BKN	Broken	5/8 - 7/8
OVC	Overcast	8/8

1. The abbreviation **CLR** shall be used at automated stations when no layers at or below 12,000 feet are reported; the abbreviation **SKC** shall be used at manual stations when no layers are reported.
2. Any layer amount less than 1/8 is reported as FEW.

The height of the base of each layer, $h_s h_s h_s$, shall be coded in hundreds of feet above the surface using three digits in accordance with Table 12-4.

Table 12-4. Increments of Reportable Values of Sky Cover Height

Range of Height Values (feet)	Reportable Increment (feet)
≤5,000	To nearest 100
>5,000 but ≤10,000	To nearest 500
>10,000	To nearest 1,000

At manual stations, cumulonimbus (**CB**) or towering cumulus (**TCU**) shall be appended to the associated layer. For example, a scattered layer of towering cumulus at 1,500 feet would be coded "SCT015TCU" and would be followed by a space if there were additional higher layers to code.

12.6.10 Temperature/Dew Point Group ($T'T'/T'_d T'_d$). The standards for observing and reporting temperature and dew point are given in Chapter 10. The temperature shall be separated from the dew point with a solidus "/".

The temperature and dew point shall be coded as two digits rounded to the nearest whole degree Celsius (see paragraph 2.6.3). For example, a temperature of 0.3°C would be coded as "00". Sub-zero temperatures and dew points shall be prefixed with an **M**. For example, a temperature of 4°C with a dew point of -2°C would be coded as "04/M02"; a temperature of -0.5°C would be coded as "M00".

If the temperature is not available, the entire temperature/dew point group shall not be coded. If the dew point is not available, the temperature shall be coded followed by a solidus "/" and no entry made for dew point. For example, a temperature of 1.5°C and a missing dew point would be coded as "02/".

12.6.11 **Altimeter** $(AP_HP_HP_HP_H)$**.** The standards for observing and reporting altimeter are described in Chapter 11.

The altimeter group always starts with an **A** (the international indicator for altimeter in inches of mercury). The altimeter shall be coded as a four digit group immediately following the **A** using the tens, units, tenths, and hundredths of inches of mercury. The decimal point is not coded.

12.7 Remarks (RMK)

Remarks shall be included in all METAR and SPECI, if appropriate.

Remarks shall be separated from the body of the report by a space and the contraction **RMK**. If there are no remarks, the contraction **RMK** is not required.

METAR/SPECI remarks fall into 2 categories: (1) Automated, Manual, and Plain Language (see paragraph 12.7.1), and (2) Additive and Maintenance Data (see paragraph 12.7.2).

Remarks shall be made in accordance with the following:

a. Where plain language is called for, authorized contractions, abbreviations, and symbols should be used to conserve time and space. However, in no case should an essential remark, of which the observer is aware, be omitted for the lack of readily available contractions. In such cases, the only requirement is that the remark be clear. For a detailed list of authorized contractions, see FAA Order 7340 Series, *Contractions*.

b. Time entries shall be made in minutes past the hour if the time reported occurs during the same hour the observation is taken. Hours and minutes shall be used if the hour is different, or this Handbook prescribes the use of the hour and minutes.

c. Present weather coded in the body of the report as **VC** may be further described, i.e., direction from the station, if known. Weather phenomena beyond 10 statute miles of the point(s) of observation shall be coded as distant (**DSNT**) followed by the direction from the station. For example, precipitation of unknown intensity within 10 statute miles east of the station would be coded as "VCSH E"; lightning 25 statute miles west of the station would be coded as "LTG DSNT W".

d. Distance remarks shall be statute miles except for automated lightning remarks which are in nautical miles.

e. Movement of clouds or weather, if known, shall be coded with respect to the direction toward which the phenomena is moving. For example, a thunderstorm moving toward the northeast would be coded as "TS MOV NE".

f. Directions shall use the eight points of the compass coded in a clockwise order.

g. Insofar as possible, remarks shall be entered in the order they are presented in the following paragraphs.

12.7.1 <u>Automated, Manual, and Plain Language Remarks.</u> These remarks generally elaborate on parameters reported in the body of the report. Automated and manual remarks may be generated either by an automated or manual station. Plain language remarks are only provided from manual stations.

  a. **Volcanic Eruptions (Plain Language).** Volcanic eruptions shall be coded.

  The remark shall be plain language and contain the following, if known:

  (1) **Name** of volcano.

  (2) **Latitude and longitude** or the direction and the approximate distance from the station.

  (3) **Date/Time** (UTC) of the eruption.

  (4) Size **description**, approximate height, and direction of movement **of the ash cloud**.

  (5) Any **other pertinent data** about the eruption.

  For example, a remark on a volcanic eruption would look like the following:

  MT. AUGUSTINE VOLCANO 70 MILES SW ERUPTED 231505 LARGE ASH CLOUD EXTENDING TO APRX 30000 FEET MOVING NE.

  Pre-eruption volcanic activity shall not be coded. Pre-eruption refers to unusual and/or increasing volcanic activity which could presage a volcanic eruption.

  b. **Funnel Cloud (Tornadic activity_B/E(hh)mm_LOC/DIR_(MOV)).** At manual stations, tornadoes, funnel clouds, or waterspouts shall be coded in the format, **Tornadic activity_B/E(hh)mm_LOC/DIR_(MOV)**, where **TORNADO, FUNNEL CLOUD,** or **WATERSPOUT** identifies the specific tornadic activity, **B/E** denotes the beginning and/or ending time, **(hh)mm** is the time of occurrence (only the minutes are required if the hour can be inferred from the report time), **LOC/DIR** is the location and/or direction of the phenomena from the station, and **MOV** is the movement, if known (see paragraphs 8.3.3.c, 8.5.3.c, and 8.5.5.b). Tornadic activity shall be coded as the first remark after the "RMK" entry. For example, "TORNADO B13 6 NE" would indicate that a tornado, which began at 13 minutes past the hour, was 6 statute miles northeast of the station.

  c. **Type of Automated Station (AO1 or AO2).** **AO1** or **AO2** shall be coded in all METAR/SPECI from automated stations. Automated stations without a precipitation discriminator shall be identified as **AO1**; automated station with a precipitation discriminator shall be identified as **AO2**.

  d. **Peak Wind (PK_WND_dddff(f)/(hh)mm).** The peak wind shall be coded in the format, **PK_WND dddff(f)/(hh)mm** of the next METAR, where **PK_WND** is the remark identifier, **ddd** is the direction of the peak wind, **ff(f)** is the peak wind speed since the last METAR, and **(hh)mm** is the time of occurrence (only the minutes are required if the hour can be inferred from the report time) (see paragraphs 5.4.5 and 5.5.5). There shall be a space between the two elements of the remark identifier and the wind direction/speed group; a solidus "/" (without spaces) shall separate the wind direction/speed group and the time. For example, a peak wind of 45 knots from 280 degrees that occurred at 15 minutes past the hour would be coded "PK WND 28045/15".

  e. **Wind Shift (WSHFT_(hh)mm).** A wind shift shall be coded in the format, **WSHFT_(hh)mm**, where **WSHFT** is the remark identifier and **(hh)mm** is the time the wind shift began (only the minutes are required if the hour can be inferred from the report time) (see paragraphs 5.4.6 and 5.5.6). The contraction **FROPA** may be entered following the time if it is reasonably certain that the wind shift was the result of a frontal passage. There shall be a space between the remark identifier and the time and, if applicable, between the time and the frontal passage contraction. For example, a remark reporting a wind shift accompanied by a frontal passage that began at 30 minutes after the hour would be coded as "WSHFT 30 FROPA".

f. **Tower or Surface Visibility (TWR_VIS_vvvvv or SFC_VIS_vvvvv).** Tower visibility or surface visibility (see paragraphs 6.5.4 and 6.5.5) shall be coded in the formats, **TWR_VIS_vvvvv** or **SFC_VIS_vvvvv,** respectively, where **vvvvv** is the observed tower/surface visibility value. A space shall be coded between each of the remark elements. For example, the control tower visibility of 1 1/2 statute miles would be coded "TWR VIS 1 1/2".

g. **Variable Prevailing Visibility (VIS_$v_nv_nv_nv_nv_n$V$v_xv_xv_xv_xv_x$).** Variable prevailing visibility shall be coded in the format **VIS_$v_nv_nv_nv_nv_n$V$v_xv_xv_xv_xv_x$,** where **VIS** is the remark identifier, $v_nv_nv_nv_nv_n$ is the lowest visibility evaluated, **V** denotes variability between two values, and $v_xv_xv_xv_xv_x$ is the highest visibility evaluated. There shall be one space following the remark identifier; no spaces between the letter **V** and the lowest/highest values. For example, a visibility that was varying between 1/2 and 2 statute miles would be coded "VIS 1/2V2" (see paragraphs 6.4.5 and 6.5.3).

h. **Sector Visibility (VIS_[DIR]_vvvvv) [Plain Language].** The sector visibility shall be coded in the format, **VIS_[DIR]_vvvvv,** where **VIS** is the remark identifier, **[DIR]** defines the sector to 8 points of the compass, and **vvvvv** is the sector visibility in statute miles, using the appropriate set of values in Table 12-1 (see paragraphs 6.4.6 and 6.5.7). For example, a visibility of 2 1/2 statute miles in the northeastern octant would be coded "VIS NE 2 1/2".

i. **Visibility At Second Location (VIS_vvvvv_[LOC]).** At designated automated stations, the visibility at a second location shall be coded in the format **VIS_vvvvv_[LOC],** where **VIS** is the remark identifier, **vvvvv** is the measured visibility value, and **[LOC]** is the specific location of the visibility sensor(s) at the station (see paragraph 6.5.6). This remark shall only be generated when the condition is lower than that contained in the body of the report. For example, a visibility of 2 1/2 statute miles measured by a second sensor located at runway 11 would be coded "VIS 2 1/2 RWY11".

j. **Lightning (Frequency_LTG(type)_[LOC]).**

    (1) When lightning is observed at a manual station, the frequency, type of lightning, and location shall be reported. The remark shall be coded in the format **Frequency_LTG(type)_[LOC].** The contractions for the type and frequency of lightning shall be based on Table 12-5. The location and direction shall be coded in accordance with paragraph 12.7.c. For example, "OCNL LTGICCG OHD", "FRQ LTG VC", or "LTG DSNT W".

    (2) When lightning is detected by an <u>automated</u> system:

        (a) Within 5 nautical miles of the Airport Location Point (ALP), it will be reported as **TS** in the body of the report with no remark;

        (b) Between 5 and 10 nautical miles of the ALP, it will be reported as **VCTS** in the body of the report with no remark;

        (c) Beyond 10 but less than 30 nautical miles of the ALP, it will be reported in remarks only as **LTG DSNT** followed by the direction from the ALP.

Table 12-5. Type and Frequency of Lightning

Type of Lightning		
Type	**Contraction**	**Definition**
Cloud-ground	CG	Lightning occurring between cloud and ground.
In-cloud	IC	Lightning which takes place within the cloud.
Cloud-cloud	CC	Streaks of lightning reaching from one cloud to another.
Cloud-air	CA	Streaks of lightning which pass from a cloud to the air, but do not strike the ground.
Frequency of Lightning		
Frequency	**Contraction**	**Definition**
Occasional	OCNL	Less than 1 flash/minute.
Frequent	FRQ	About 1 to 6 flashes/minute.
Continuous	CONS	More than 6 flashes/minute.

k. **Beginning and Ending of Precipitation (w'w'B(hh)mmE(hh)mm).** At designated stations, the beginning and ending of precipitation shall be coded in the format, **w'w'B(hh)mmE(hh)mm**, where **w'w'** is the type of precipitation, **B** denotes the beginning, **E** denotes the ending, and **(hh)mm** is the time of occurrence (only the minutes are required if the hour can be inferred from the report time) (see paragraph 8.5.5.a). There shall be no spaces between the elements. The coded remarks are not required in SPECI and should be reported in the next METAR. Intensity qualifiers shall not be coded. For example, if rain began at 0005, ended at 0030, and snow began at 0020, and ended at 0055, the remarks would be coded "RAB05E30SNB20E55". If the precipitation were showery, the remark would be coded "SHRAB05E30SHSNB20E55".

l. **Beginning and Ending of Thunderstorms (TSB(hh)mmE(hh)mm).** The beginning and ending of thunderstorm(s) shall be coded in the format, **TSB(hh)mmE(hh)mm**, where **TS** indicates thunderstorm, **B** denotes the beginning, **E** denotes the ending, and **(hh)mm** is the time of occurrence (only the minutes are required if the hour can be inferred from the report time) (see paragraph 8.5.4). There shall be no spaces between the elements. For example, if a thunderstorm began at 0159 and ended at 0230, the remark would be coded "TSB0159E30".

m. **Thunderstorm Location (TS_LOC_(MOV_DIR)) [Plain Language].** Thunderstorm(s) shall be coded in the format, **TS_LOC_(MOV_DIR)**, where **TS** identifies the thunderstorm activity, **LOC** is the location of the thunderstorm(s) from the station, and **MOV_DIR** is the movement with direction, if known (see paragraph 8.4.1.b(5) and 8.5.4). For example, a thunderstorm southeast of the station and moving toward the northeast would be coded "TS SE MOV NE".

n. **Hailstone Size (GR_[size]) [Plain Language].** At designated stations, the hailstone size shall be coded in the format, **GR_[size]**, where **GR** is the remark identifier and **[size]** is the diameter of the largest hailstone. The hailstone size shall be coded in 1/4 inch increments (see paragraph 8.5.1.c(8)). For example, "GR 1 3/4" would indicate that the largest hailstones were 1 3/4 inches in diameter. If **GS** is coded in the body of the report, no hailstone size remark is required.

o. **Virga (VIRGA_(DIR)) [Plain Language].** Virga shall be coded in the format, **VIRGA_(DIR)**, where **VIRGA** is the remark identifier and **DIR** is the direction from the station. The direction of the phenomena from the station is optional, e.g., "VIRGA" or "VIRGA SW".

p. **Variable Ceiling Height (CIG_$h_nh_nh_n$V$h_xh_xh_x$).** The variable ceiling height shall be coded in the format, **CIG_$h_nh_nh_n$V$h_xh_xh_x$**, where **CIG** is the remark identifier, $h_nh_nh_n$ is the lowest ceiling height evaluated, **V** denotes variability between two values, and $h_xh_xh_x$ is the highest ceiling height evaluated (see paragraph 9.5.7 and Table 9-1). There shall be one space following the remark identifier; no spaces between the letter **V** and the lowest/highest ceiling values. For example, "CIG 005V010" would indicate a ceiling that was varying between 500 and 1,000 feet.

q. **Obscurations (w'w'_[$N_sN_sN_s$]$h_sh_sh_s$).** **[Plain Language]** Obscurations (surface-based or aloft) shall be coded in the format, **w'w'_[$N_sN_sN_s$]$h_sh_sh_s$**, where **w'w'** is the weather causing the obscuration at the surface or aloft, $N_sN_sN_s$ is the applicable sky cover amount of the obscuration aloft (FEW, SCT, BKN, OVC) or at the surface (FEW, SCT, BKN), and $h_sh_sh_s$ is the applicable height (see paragraphs 9.4.3 and 9.5.6). Surface-based obscurations shall have a height of "000". There shall be a space separating the weather causing the obscuration and the sky cover amount; there shall be no space between the sky cover amount and the height. For example, fog hiding 3-4 oktas of the sky would be coded "FG SCT000"; a broken layer at 2,000 feet composed of smoke would be coded "FU BKN020".

r. **Variable Sky Condition ($N_sN_sN_s$($h_sh_sh_s$)_V_$N_sN_sN_s$).** **[Plain Language]** The variable sky condition remark shall be coded in the format, **$N_sN_sN_s$($h_sh_sh_s$)_V_$N_sN_sN_s$**, where **$N_sN_sN_s$($h_sh_sh_s$)** and **$N_sN_sN_s$** identifies the two operationally significant sky conditions and **V** denotes the variability between the two ranges (see paragraphs 9.4.2.d and 9.5.9). If there are several layers with the same sky condition amount, the layer height (**$h_sh_sh_s$**) of the variable layer shall be coded. For example, a cloud layer at 1,400 feet that is varying between broken and overcast would be coded "BKN014 V OVC".

s. **Significant Cloud Types [Plain Language].** The significant cloud type remark shall be coded in all reports in the following manner (see paragraphs 9.4.6 and 9.5.10):

(1) **Cumulonimbus or Cumulonimbus Mammatus (CB or CBMAM_LOC_(MOV_DIR).** Cumulonimbus or cumulonimbus mammatus, as appropriate, (for which no thunderstorm is being reported) shall be coded in the format, **CB or CBMAM_LOC_(MOV_DIR)**, where **CB** or **CBMAM** is the cloud type, **LOC** is the direction from the station, and **MOV_DIR** is the movement with direction (if known). The cloud type, location, movement, and direction entries shall be separated from each other with a space. For example, a CB up to 10 statute miles west of the station moving toward the east would be coded "CB W MOV E". If the CB was more than 10 statute miles to the west, the remark would be coded "CB DSNT W".

(2) **Towering cumulus (TCU_[DIR]).** Towering cumulus clouds shall be coded in the format, **TCU_[DIR]**, where **TCU** is the cloud type and **DIR** is the direction from the station. The cloud type and direction entries shall be separated by a space. For example, a towering cumulus cloud up to 10 statute miles west of the station would be coded "TCU W".

(3) **Altocumulus castellanus (ACC_[DIR]).** Altocumulus castellanus shall be coded in the format, **ACC_[DIR]**, where **ACC** is the cloud type and **DIR** is the direction from the station. The cloud type and direction entries shall be separated by a space. For example, an altocumulus cloud 5 to 10 statute miles northwest of the station would be coded "ACC NW".

(4) **Standing lenticular or Rotor clouds (CLD_[DIR]).** Stratocumulus (SCSL), altocumulus (ACSL), or cirrocumulus (CCSL), or rotor clouds shall be coded in the format, **CLD_[DIR]**, where **CLD** is the cloud type and **DIR** is the direction from the station. The cloud type and direction entries shall be separated by a space. For example, altocumulus standing lenticular clouds observed southwest through west of the station would be coded "ACSL SW−W"; an apparent rotor cloud 5 to 10 statute miles northeast of the station would be coded "APRNT ROTOR CLD NE"; and cirrocumulus clouds south of the station would be coded "CCSL S".

t. **Ceiling Height at Second Location (CIG_hhh_[LOC]).** At designated stations, the ceiling height at a second location shall be coded in the format, **CIG_hhh_[LOC]**, where **CIG** is the remark identifier, **hhh** is the measured height of the ceiling, and **[LOC]** is the specific location of the ceilometer(s) at the station (see paragraph 9.5.8). This remark shall only be generated when the ceiling is lower than that contained in the body of the report. For example, if the ceiling measured by a second sensor located at runway 11 is broken at 200 feet, the remark would be "CIG 002 RWY11".

u. **Pressure Rising or Falling Rapidly (PRESRR/PRESFR).** At designated stations, when the pressure is rising or falling rapidly at the time of observation (see paragraphs 11.4.6 and 11.5.5), the remark **PRESRR** (pressure rising rapidly) or **PRESFR** (pressure falling rapidly) shall be included in the report.

v. **Sea-Level Pressure (SLPppp).** At designated stations, the sea-level pressure shall be coded in the format **SLPppp**, where **SLP** is the remark identifier and **ppp** is the sea-level pressure in hectopascals (see paragraphs 11.4.4 and 11.5.4). For example, a sea-level pressure of 998.2 hectopascals would be coded as "SLP982". For a METAR, if sea-level pressure is not available, it is coded as "**SLPNO**".

w. **Aircraft Mishap (ACFT_MSHP) [Plain Language].** If a report is taken to document weather conditions when notified of an aircraft mishap, the remark **ACFT_MSHP** shall be coded in the report but not transmitted. The act of non-transmission shall be indicated by enclosing the remark in parentheses in the record, i.e., "(ACFT MSHP)".

x. **No SPECI Reports Taken (NOSPECI) [Plain Language].** At manual stations where SPECI's are not taken, the remark **NOSPECI** shall be coded to indicate that no changes in weather conditions will be reported until the next METAR.

y. **Snow Increasing Rapidly (SNINCR_[inches-hour/inches on ground]).** At designated stations, the snow increasing rapidly remark shall be reported, in the next METAR, whenever the snow depth increases by 1 inch or more in the past hour. The remark shall be coded in the format, **SNINCR [inches-hour/inches on ground]**, where **SNINCR** is the remark indicator, **inches-hour** is the depth increase in the past hour, and **inches on ground** is the total depth of snow on the ground at the time of the report. The depth increase in the past hour and the total depth on the ground are separated from each other by a solidus "/". For example, a snow depth increase of 2 inches in the past hour with a total depth on the ground of 10 inches would be coded "SNINCR 2/10".

z. **Other Significant Information [Plain Language].** Agencies may add to a report other information significant to their operations, such as information on fog dispersal operations, runway conditions, "FIRST" or "LAST" report from station, etc.

12.7.2 **Additive and Automated Maintenance Data.** Additive data groups are only reported at designated stations. The maintenance data groups are only reported from automated stations.

    a. **Precipitation**

        (1) **Amount of Precipitation.** The amount of liquid precipitation shall be coded as the depth of precipitation that accumulates in an exposed vessel during the time period being evaluated. The amount of freezing or frozen precipitation shall be the water equivalent of the solid precipitation accumulated during the appropriate time period.

        (2) **Units of Measure for Precipitation.** Precipitation measurements shall be in inches, tenths of inches, or hundredths of inches depending on the precipitation being measured (see Table 12-6).

Table 12.6. Units of Measure for Precipitation

Type of Measurement	Unit of Measure
Liquid Precipitation	0.01 inch
Water Equivalent of Solid Precipitation	0.01 inch
Solid Precipitation	0.1 inch
Snow Depth	1.0 inch

        (3) **Depth of Freezing or Frozen Precipitation.** The depth of freezing and/or frozen precipitation shall be the actual vertical depth of the precipitation accumulated on a horizontal surface during the appropriate time period (see paragraphs 12.7.2.a(3)(b) and 12.7.2.a(3)(c). If snow falls, melts, and refreezes, the depth of ice formed shall be included in the measurement.

            (a) **Hourly Precipitation Amount (Prrrr).** At designated automated stations, the hourly precipitation amount shall be coded in the format, **Prrrr**, where **P** is the group indicator and **rrrr** is the water equivalent of all precipitation that has occurred since the last METAR (see paragraph 12.7.2.a(1)). The amount shall be coded in hundredths of an inch. For example, "P0009" would indicate 9/100 of an inch of precipitation fell in the past hour; "P0000" would indicate that less than 1/100 of an inch of precipitation fell in the past hour.

            The group shall be omitted if no precipitation occurred since the last METAR.

            (b) **3- and 6-Hour Precipitation Amount (6RRRR).** At designated stations, the 3- and 6-hourly precipitation group shall be coded in the format, **6RRRR**, where **6** is the group indicator and **RRRR** is the amount of precipitation. The amount of precipitation (water equivalent) accumulated in the past 3 hours shall be reported in the 3-hourly report; the amount accumulated in the past 6 hours shall be reported in the 6-hourly report. The amount of precipitation shall be coded in inches, using the tens, units, tenths and hundredths digits of the amount. When an indeterminable amount of precipitation has occurred during the period, **RRRR** shall be coded 6////. For example, 2.17 inches of precipitation would be coded "60217". A trace shall be coded "60000".

            (c) **24-Hour Precipitation Amount ($7R_{24}R_{24}R_{24}R_{24}$).** At designated stations, the 24-hour precipitation amount shall be coded in the format, $7R_{24}R_{24}R_{24}R_{24}$, where **7** is the group indicator and $R_{24}R_{24}R_{24}R_{24}$ is the 24-hour precipitation amount. The 24-hour precipitation amount shall be included in the 1200 UTC (or other agency designated time) report whenever more than a trace of precipitation (water equivalent) has fallen in the preceding 24 hours. The amount of precipitation shall be coded by using the tens, units, tenths, and hundredths of inches (water equivalent) for the 24-hour period. If more than a trace (water equivalent) has occurred and the amount cannot be determined, the group shall be coded

7/////. For example, 1.25 inches of precipitation (water equivalent) in the past 24 hours shall be coded "70125".

    (d)   **Snow Depth on Ground (4/sss).** At designated stations, the total snow depth on the ground group shall be coded in the 0000, 0600, 1200, and 1800 UTC observations whenever there is more than a trace of snow on the ground. The remark shall be coded in the format, **4/sss**, where **4/** is the group indicator and **sss** is the snow depth in whole inches using three digits. For example, a snow depth of 21 inches shall be coded as "4/021".

    (e)   **Water Equivalent of Snow on Ground (933RRR).** At designated stations, the water equivalent of snow on the ground shall be coded each day, in the 1800 UTC report, if the average snow depth is 2 inches or more. The remark shall be coded in the format, **933RRR**, where **933** is the group indicator and **RRR** is the water equivalent of snow, i.e., snow, snow pellets, snow grains, ice pellets, ice crystals, and hail, on the ground. The water equivalent shall be coded in tens, units, and tenths of inches, using three digits. If the water equivalent of snow consists entirely of hail, the group shall not be coded. A water equivalent of 3.6 inches of snow would be coded as "933036"; a water equivalent of 12.5 would be coded as "933125". This value is never estimated, ratios (e.g., 10 to 1) or temperature/snow water equivalent tables are not to be used to determine water equivalency of snow for this group.

b.   **Cloud Types (8/$C_L C_M C_H$).** At designated stations, the group, **8/$C_L C_M C_H$**, shall be reported and coded in 3- and 6-hourly reports when clouds are observed. The predominant low cloud (C_L), middle cloud (C_M), and high cloud (C_H), shall be identified in accordance with the WMO *International Cloud Atlas*, Volumes I and II, or the WMO *Abridged International Cloud Atlas* or agency observing aids for cloud identification. A **0** shall be coded for the low, middle, or high cloud type if no cloud is present in that classification. A solidus "/" shall be coded for layers above an overcast. If no clouds are observed due to clear skies, the cloud type group shall not be coded. For example, a report of "8/6//" would indicate an overcast layer of stratus clouds; a report of "8/903" would indicate cumulonimbus type low clouds, no middle clouds, and dense cirrus high clouds.

c.   **Duration of Sunshine (98mmm).** The duration of sunshine that occurred the previous calendar day shall be coded in the 0800 UTC report. If the station is closed at 0800 UTC, the group shall be coded in the first 6-hourly METAR after the station opens. The duration of sunshine shall be coded in the format, **98mmm**, where **98** is the group indicator and **mmm** is the total minutes of sunshine. The minutes of sunshine shall be coded by using the hundreds, tens, and units digits. For example, 96 minutes of sunshine would be coded "98096". If no sunshine occurred, the group would be coded "98000".

d.   **Hourly Temperature and Dew Point ($Ts_n T'T'T's_n T'_d T'_d T'_d$).** At designated stations, the hourly temperature and dew point group shall be coded to the tenth of a degree Celsius in the format, **$Ts_n T'T'T's_n T'_d T'_d T'_d$**, where **T** is the group indicator, **s_n** is the sign of the temperature, **T'T'T'** is the temperature, and **$T'_d T'_d T'_d$** is the dew point (see paragraphs 10.5.1 and 10.5.3). The sign of the temperature and dew point shall be coded as 1 if the value is below 0°C and 0 if the value is 0°C or higher. The temperature and dew point shall be reported in tens, units, and tenths of degrees Celsius. There shall be no spaces between the entries. For example, a temperature of 2.6°C and dew point of -1.5°C would be reported in the body of the report as "03/M01" and the **$Ts_n T'T'T's_n T'_d T'_d T'_d$** group as "T00261015". If dew point is missing report the temperature; if the temperature is missing do not report the temperature/dew point group.

e.   **6-Hourly Maximum Temperature ($1s_nT_xT_xT_x$).** At designated stations, the 6-hourly maximum temperature group shall be coded in the format, $1s_nT_xT_xT_x$, where **1** is the group indicator, s_n is the sign of the temperature, $T_xT_xT_x$ is the maximum temperature in tenths of degrees Celsius using three digits (see paragraphs 10.4.4, 10.5.2, and 10.5.3). The sign of the maximum temperature shall be coded as 1 if the maximum temperature is below 0°C and 0 if the maximum temperature is 0°C or higher. For example, a maximum temperature of −2.1°C would be coded "11021"; 14.2°C would be coded "10142".

f.   **6-Hourly Minimum Temperature ($2s_nT_nT_nT_n$).** At designated stations, the 6-hourly minimum temperature group shall be coded in the format, $2s_nT_nT_nT_n$, where **2** is the group indicator, s_n is the sign of the temperature, and $T_nT_nT_n$ is the minimum temperature in tenths of degrees Celsius using three digits (see paragraphs 10.4.4, 10.5.2, and 10.5.3). The sign of the minimum temperature shall be coded as 1 if the minimum temperature is below 0°C and 0 if the minimum temperature is 0°C or higher. For example, a minimum temperature of −0.1°C would be coded "21001"; 1.2°C would be coded "20012".

g.   **24-Hour Maximum and Minimum Temperature ($4s_nT_xT_xT_xs_nT_nT_nT_n$).** At designated stations, the 24-hour maximum temperature and the 24-hour minimum temperature shall be coded in the format, $4s_nT_xT_xT_xs_nT_nT_nT_n$, where **4** is the group indicator, s_n is the sign of the temperature, $T_xT_xT_x$ is the maximum 24-hour temperature, and $T_nT_nT_n$ is the 24-hour minimum temperature (see paragraphs 10.4.4, 10.5.2, and 10.5.3). $T_xT_xT_x$ and $T_nT_nT_n$ shall be coded in tenths of degrees Celsius using three digits. The sign of the maximum or minimum temperature shall be coded as 1 if it is below 0°C and 0 if it is 0°C or higher. For example, a 24-hour maximum temperature of 10.0°C and a 24-hour minimum temperature of −1.5°C would be coded "401001015"; a 24-hour maximum temperature of 11.2°C and a 24-hour minimum temperature of 8.4°C would be coded as "401120084".

h.   **3-Hourly Pressure Tendency (5appp).** At designated stations, the 3-hourly pressure tendency group shall be coded in the format, **5appp**, where **5** is the group indicator, **a** is the character of pressure change over the past 3 hours (see Table 12-7), and **ppp** is the amount of barometric change in tenths of hectopascals (see Table 12-8). The amount of barometric change shall be coded using the tens, units, and tenths digits (see paragraphs 11.4.7 and 11.5.4). For example, a steady increase of 3.2 hectopascals in the past three hours would be coded "52032".

Table 12-7 Characteristics of Barometer Tendency

Primary Requirement	Description	Code Figure
Atmospheric pressure now higher than 3 hours ago.	Increasing, then decreasing.	0
	Increasing, then steady, or increasing then increasing more slowly.	1
	Increasing steadily or unsteadily.	2
	Decreasing or steady, then increasing; or increasing then increasing more rapidly.	3
Atmospheric pressure now same as 3 hours ago.	Increasing, then decreasing.	0
	Steady.	4
	Decreasing, then increasing.	5
Atmospheric pressure now lower than 3 hours ago.	Decreasing, then increasing.	5
	Decreasing then steady; or decreasing then decreasing more slowly.	6
	Decreasing steadily or unsteadily.	7
	Steady or increasing, then decreasing; or decreasing then decreasing more rapidly.	8

Table 12-8. 3-Hour Pressure Change

Amount of Barometric Change (Rise or Fall) in the Past 3 Hours "ppp"								
Code Figure	Inches of Mercury	Hectopascals	Code Figure	Inches of Mercury	Hectopascals	Code Figure	Inches of Mercury	Hectopascals
000	0.000	0.0	068	0.200	6.8	135	0.400	13.5
002	0.005	0.2	069	0.205	6.9	137	0.405	13.7
003	0.010	0.3	071	0.210	7.1	139	0.410	13.9
005	0.015	0.5	073	0.215	7.3	141	0.415	14.1
007	0.020	0.7	075	0.220	7.5	142	0.420	14.2
008	0.025	0.8	076	0.225	7.6	144	0.425	14.4
010	0.030	1.0	078	0.230	7.8	146	0.430	14.6
012	0.035	1.2	080	0.235	8.0	147	0.435	14.7
014	0.040	1.4	081	0.240	8.1	149	0.440	14.9
015	0.045	1.5	083	0.245	8.3	151	0.445	15.1
017	0.050	1.7	085	0.250	8.5	152	0.450	15.2
019	0.055	1.9	086	0.255	8.6	154	0.455	15.4
020	0.060	2.0	088	0.260	8.8	156	0.460	15.6
022	0.065	2.2	090	0.265	9.0	157	0.465	15.7
024	0.070	2.4	091	0.270	9.1	159	0.470	15.9
025	0.075	2.5	093	0.275	9.3	161	0.475	16.1
027	0.080	2.7	095	0.280	9.5	163	0.480	16.3
029	0.085	2.9	097	0.285	9.7	164	0.485	16.4
030	0.090	3.0	098	0.290	9.8	166	0.490	16.6
032	0.095	3.2	100	0.295	10.0	168	0.495	16.8
034	0.100	3.4	102	0.300	10.2	169	0.500	16.9
036	0.105	3.6	103	0.305	10.3	171	0.505	17.1
037	0.110	3.7	105	0.310	10.5	173	0.510	17.3
039	0.115	3.9	107	0.315	10.7	174	0.515	17.4
041	0.120	4.1	108	0.320	10.8	176	0.520	17.6
042	0.125	4.2	110	0.325	11.0	178	0.525	17.8
044	0.130	4.4	112	0.330	11.2	179	0.530	17.9
046	0.135	4.6	113	0.335	11.3	181	0.535	18.1
047	0.140	4.7	115	0.340	11.5	183	0.540	18.3
049	0.145	4.9	117	0.345	11.7	185	0.545	18.5
051	0.150	5.1	119	0.350	11.9	186	0.550	18.6
052	0.155	5.2	120	0.355	12.0	188	0.555	18.8
054	0.160	5.4	122	0.360	12.2	190	0.560	19.0
056	0.165	5.6	124	0.365	12.4	191	0.565	19.1
058	0.170	5.8	125	0.370	12.5	193	0.570	19.3
059	0.175	5.9	127	0.375	12.7	195	0.575	19.5
061	0.180	6.1	129	0.380	12.9	196	0.580	19.6
063	0.185	6.3	130	0.385	13.0	198	0.585	19.8
064	0.190	6.4	132	0.390	13.2	200	0.590	20.0
066	0.195	6.6	134	0.395	13.4	201	0.595	20.1

i. **Sensor Status Indicators.** Sensor status indicators should be reported as indicated below:

(1) if the Runway Visual Range should be reported but is missing, **RVRNO** shall be coded.

(2) when automated stations are equipped with a present weather identifier and that sensor is not operating, the remark **PWINO** shall be coded.

(3) when automated stations are equipped with a tipping bucket rain gauge and that sensor is not operating, **PNO** shall be coded.

(4) when automated stations are equipped with a freezing rain sensor and that sensor is not operating, the remark **FZRANO** shall be coded.

(5)   when automated stations are equipped with a lightning detection system and that sensor is not operating, the remark **TSNO** shall be coded.

(6)   when automated stations are equipped with a secondary visibility sensor and that sensor is not operating, the remark **VISNO_LOC** shall be coded.

(7)   when automated stations are equipped with a secondary ceiling height indicator and that sensor is not operating, the remark **CHINO_LOC** shall be coded.

j.   **Maintenance Indicator.** A maintenance indicator sign, **$**, shall be coded when an automated system detects that maintenance is needed on the system.

APPENDIX A

GLOSSARY[1]

---

[1]The terms and definitions presented in this glossary are in accordance with their usage in this Handbook.

3-hourly report. A METAR taken at 0300, 0900, 1500, or 2100 UTC.

6-hourly report. A METAR taken at 0000, 0600, 1200, or 1800 UTC.

actual time of observation. For METARs, it is the time the last element of the report is observed or evaluated. For SPECIs, it is the time that the criteria for a SPECI was met or noted.

additive data. A group of coded remarks that includes pressure tendency, amount of precipitation, and maximum/minimum temperature during specified periods of time.

aircraft mishap. An inclusive term to denote the occurrence of an aircraft accident or incident.

Airport Location Point. ALP, the permanent airport reference point defined by the latitude and longitude published in the Airport Facility Directory.

algorithm. A set of rules implemented (usually in a computer) to process data and generate defined outputs.

altimeter setting. That pressure value to which an aircraft altimeter scale is set so that it will indicate the altitude above mean sea-level of an aircraft on the ground at the location for which the value was determined.

archive. A permanent record of surface weather reports and related data used to establish a climatological record for the United States.

atmospheric pressure. The pressure exerted by the atmosphere at a given point (see altimeter setting, pressure, sea-level pressure, station pressure).

augmented report. A meteorological report prepared by an automated surface weather observing system for transmission with certified observers signed on to the system to add information to the report.

automated report. A meteorological report prepared by an automated surface weather observing system for transmission, and with no certified weather observers signed on to the system.

backup. An alternate method for providing a meteorological report, parts of reports or documentation of reports when the primary method is unavailable.

barogram. An analog record of pressure produced by a barograph.

barograph. A recording barometer.

barometer. An instrument that measures atmospheric pressure.

barometric pressure. The actual pressure value indicated by a pressure sensor.

blowing. A descriptor used to amplify observed weather phenomena whenever the phenomena are raised to a height of 6 feet or more above the ground.

blowing dust. Dust picked up locally from the surface of the earth and blown about in clouds or sheets, reducing the reported horizontal visibility to less than 7 statute miles.

blowing sand. Sand particles picked up from the surface of the earth by the wind to moderate heights above the ground, reducing the reported horizontal visibility to less than 7 statute miles.

blowing snow. Snow lifted from the surface of the earth by the wind to a height of 6 feet or more above the ground and blown about in such quantities that the reported horizontal visibility is reduced to less than 7 miles.

blowing spray.   Water droplets torn by the wind from a body of water, generally from the crests of waves, and carried up into the air in such quantities that they reduce the reported horizontal visibility to less than 7 statute miles.

body of report.   That portion of a METAR or SPECI beginning with the type of report and ending with the altimeter setting.

broken layer.   A layer covering whose summation amount of sky cover is 5/8ths through 7/8ths.

calm.   A condition when no motion of the air is detected.

candela.   A unit of luminous intensity, equal to 1/60 of the luminous intensity of a square centimeter of a black body heated to 1773.5 degrees Celsius.

ceiling.   The lowest layer aloft reported as broken or overcast; or the vertical visibility into an indefinite ceiling.

ceiling light.   A type of cloud-height indicator that uses a focused light to project vertically a narrow beam of light onto a cloud base.

ceilometer.   A device used to evaluate the height of clouds or the vertical visibility into a surface-based obscuration.

certified observer.   An individual approved by designated Federal agencies to take surface observations used in aircraft operations.

clear sky.   The absence of sky cover.

cloud.   A visible aggregate of minute water droplets or ice particles in the atmosphere above the Earth's surface.

cloud-air lightning (CA).   Streaks of lightning which pass from a cloud to the air, but do not strike the ground.

cloud-cloud lightning (CC).   Streaks of lightning reaching from one cloud to another.

cloud-ground lightning (CG):   Lightning occurring between cloud and ground.

cloud height.   The height of the base of a cloud or cloud layer above the surface of the earth.

cloud layer.   An array of clouds whose bases are at approximately the same level.

cloud movement.   The direction toward which a cloud is moving.

cloud type.   A cloud form which is identified according to the WMO International Cloud Atlas.

contraction.   A shortened form of a word, title, or phrase used for brevity.

Coordinated Universal Time (UTC).   The time in the zero degree meridian time zone.

cumulus.   A principal cloud type in the form of individual, detached elements which are generally dense and possess sharp non-fibrous outlines.

cumulonimbus.   An exceptionally dense and vertically developed cloud, occurring either isolated or as a line or wall of clouds with separated upper portions. These clouds appear as mountains or huge towers, at least a part of the upper portions of which are usually smooth, fibrous, or striated, and almost flattened.

designated RVR runway.   A runway at civilian airports designated by the FAA for reporting RVR.

designated stations. Weather observing stations that have the capability and have been instructed by their responsible agency to perform a specified task that is not required by standards to be performed at all stations.

dew point. The temperature to which a given parcel of air must be cooled at constant pressure and constant water-vapor content in order for saturation to occur.

diamond dust. See ice crystals.

dissemination. The act of delivering a completed weather report to users.

drizzle. Fairly uniform precipitation composed exclusively of fine drops (diameter less than 0.02 inch or 0.5 mm) very close together. Drizzle appears to float while following air current, although unlike fog droplets, it falls to the ground.

duration of sunshine. The amount of time sunlight was detected at a given point.

dust. (see widespread dust).

duststorm. A severe weather condition characterized by strong winds and dust-filled air over an extensive area.

element. One of the basic conditions of the atmosphere discussed in this FMH (wind, visibility, runway visual range, weather, obscurations, sky condition, temperature and dewpoint, and pressure). See parameter.

few. A layer whose summation amount of sky cover is 1/8th through 2/8ths.

field elevation. The elevation above sea level of the highest point on any of the runways of the airport.

fog. A visible aggregate of minute water particles (droplets) which are based at the Earth's surface and reduce horizontal visibility to less than 5/8 statute mile and, unlike drizzle, it does not fall to the ground.

freezing. A descriptor, FZ, used to describe drizzle and/or rain that freezes on contact with the ground or exposed objects, and used also to describe fog that is composed of minute ice crystals.

freezing drizzle. Drizzle that freezes upon impact with the ground, or other exposed objects.

freezing fog. A suspension of numerous minute ice crystals in the air, or water droplets at temperatures below 0° Celsius, based at the Earth's surface, which reduces horizontal visibility.

freezing precipitation. Any form of precipitation that freezes upon impact and forms a glaze on the ground or exposed objects.

freezing rain. Rain that freezes upon impact and forms a glaze on the ground or exposed objects.

frozen precipitation. Any form of precipitation that reaches the ground in solid form (snow, small hail and/or snow pellets, snow grains, hail, ice pellets, and ice crystals).

funnel cloud. A violent, rotating column of air which does not touch the surface, usually appended to a cumulonimbus cloud.

glaze. Ice formed by freezing precipitation covering the ground or exposed objects.

ground elevation. The official height of a weather station with reference to sea-level when a field elevation has not been established. It is the height of the ground at the base of the ceilometer.

ground fog. See shallow fog.

gust. Rapid fluctuations in wind speed with a variation of 10 knots or more between peaks and lulls.

hail. Precipitation in the form of small balls or other pieces of ice falling separately or frozen together in irregular lumps.

haze. A suspension in the air of extremely small, dry particles invisible to the naked eye and sufficiently numerous to give the air an opalescent appearance.

hectopascal. A unit of measure of atmospheric pressure equal to 100 newtons per square meter.

horizon. The actual lower boundary of the observed sky or the upper outline of terrestrial objects, including nearby natural obstructions. It is the distant line along which the earth, or the water surface at sea, and the sky appear to meet.

ice crystals (diamond dust). A fall of non-branched (snow crystals are branched) ice crystals in the form of needles, columns, or plates.

ice fog. See freezing fog.

ice pellets. Precipitation of transparent or translucent pellets of ice, which are round or irregular, rarely conical, and which have a diameter of 0.2 inch (5 mm), or less. There are two main types:

    a.   Hard grains of ice consisting of frozen raindrops, or largely melted and refrozen snowflakes.

    b.   Pellets of snow encased in a thin layer of ice which have formed from the freezing, either of droplets intercepted by the pellets, or of water resulting from the partial melting of the pellets.

in-cloud lightning (IC). Lightning which takes place within the cloud.

indefinite ceiling. The ceiling classification applied when the reported ceiling value represents the vertical visibility upward into surface-based obscuration.

intensity qualifier. Intensity qualifiers are used to describe whether a phenomena is light (-), moderate (no symbol used), or heavy (+).

layer. An array of clouds and/or obscurations whose bases are at approximately the same level.

layer amount. The amount of sky covered by clouds and/or obscurations at a given level above the Earth's surface.

layer height. The height of the bases of each reported layer of clouds and/or obscuration; or the vertical visibility into an indefinite ceiling.

lightning. The luminous phenomenon accompanying a sudden electrical discharge (see cloud-air lightning, cloud-cloud lightning, cloud-ground lightning and in-cloud lightning).

liquid precipitation. Any form of precipitation that does not fall as frozen precipitation and does not freeze upon impact.

local dissemination. The transmission or delivery of a weather report to individuals or groups of users near the observing location.

Local Standard Time (LST): A time based on the geographic location of the station in one of the legally established time zones of the globe.

long-line dissemination (also long-line transmission). The transmission of a weather report by a communication media to a group of users on a regional or national scale.

long-term retention. Retention of data for 5 years to satisfy requirements for local studies and to support litigation.

low drifting. A descriptor, DR, used to describe snow, sand, or dust raised to a height of less than 6 feet above the ground.

low drifting dust. Dust that is raised by the wind to less than 6 feet above the ground; visibility is not reduced below 7 statute miles at eye level although objects below this level may be veiled or hidden by the particles moving nearly horizontal to the ground.

low drifting sand. Sand that is raised by the wind to less than 6 feet above the ground; visibility is not reduced below 7 statute miles at eye level although objects below this level may be veiled or hidden by the particles moving nearly horizontal to the ground.

low drifting snow. Snow that is raised by the wind to less than 6 feet above the ground; visibility is not reduced below 7 statute miles at eye level although objects below this level may be veiled or hidden by the particles moving nearly horizontal to the ground.

manual station. A station, with or without an automated surface weather observing system, where the certified observers are totally responsible for all meteorological reports that are transmitted.

maximum temperature. The highest temperature during a specified time period.

may. A term used to indicate that a standard is optional.

METAR/SPECI. An evaluation of select weather elements from a point or points on or near the ground according to a set of procedures. It may include type of report, station identifier, date and time of report, a report modifier, wind, visibility, runway visual range, weather and obstructions to vision, sky condition, temperature and dew point, altimeter setting, and Remarks.

METAR/SPECI code. WMO code forms (FM 15-X Ext. METAR and FM 16-X Ext. SPECI) consisting of abbreviations, contractions, numbers, plain language, and symbols to provide a uniform means of disseminating surface weather reports.

minimum temperature. The lowest temperature during a specified time period.

mist. A visible aggregate of minute water droplets or ice crystals suspended in the atmosphere that reduces visibility to less than 7 statute miles but greater than or equal to 5/8 statute mile.

non-uniform sky condition. A localized sky condition which varies from that reported in the body of the report.

non-uniform visibility. A localized visibility which varies from that reported in the body of the report.

obscured sky. The condition when the entire sky is hidden by surface-based obscurations.

obscurations. Any phenomenon in the atmosphere, other than precipitation, that reduces the horizontal visibility in the atmosphere.

observing location. The point or points from which an element is evaluated.

observing station. The point or points from which the various elements of the report are evaluated.

overcast. A layer whose summation amount of sky cover is 8/8ths.

parameter. A subset of the group of evaluations that constitute each element of an observation; e.g., sky condition is an element, sky cover and ceiling are parameters.

partial. A descriptor, PR, used only to report fog that covers part of the airport.

partial fog. A substantial part of the station covered by fog while the remainder is clear of fog.

patches. A descriptor, BC, used only to report fog that occurs in patches at the airport.

patches (of) fog. Fog patches which randomly cover the station.

peak wind speed. The maximum instantaneous wind speed since the last METAR that exceeded 25 knots.

precipitation. Any of the forms of water particles, whether liquid or solid, that fall from the atmosphere and reach the ground.

precipitation discriminator. A sensor, or array of sensors, that differentiates between different types of precipitation (liquid, freezing, frozen).

precipitation intensity. An indication of the rate at which precipitation is falling at the time of observation.

precipitation rate. The amount of water, liquid or solid, that reaches the ground in a specified period of time.

pressure. The force exerted by a column of air above the point of measurement.

pressure change. The net difference between pressure readings at the beginning and ending of a specified interval of time.

pressure characteristic. The indication of how the pressure has been changing during a specified period of time, usually the 3-hour period preceding an observation; e.g., decreasing then increasing, pressure same or lower than 3 hours ago.

pressure falling rapidly. A decrease in station pressure at a rate of 0.06 inch of mercury or more per hour which totals 0.02 inch or more.

pressure reduction calculator. A device used to compute sea-level pressure, station pressure, altimeter setting, pressure altitude, etc.

pressure rising rapidly. An increase in station pressure at a rate of 0.06 inch of mercury or more per hour which totals 0.02 inch or more.

pressure tendency. The character and amount of atmospheric pressure change during a specified period of time, usually the 3-hour period preceding an observation.

pressure unsteady. A pressure that fluctuates by 0.03 inch of mercury or more from the mean pressure during the period of measurement.

prevailing visibility. The visibility that is considered representative of conditions at the station; the greatest distance that can be seen throughout at least half the horizon circle, not necessarily continuous.

rain. Precipitation, either in the form of drops larger than 0.02 inch (0.5 mm), or smaller drops, which in contrast to drizzle, are widely separated; for automated stations, precipitation that remains in the liquid state upon impact with the ground or other exposed objects.

Remarks. Plain language or coded data added to the body of the METAR/SPECI to report significant information not provided for in the body of the report.

rotor cloud. A turbulent cloud formation found in the lee of some large mountain barriers. The air in the cloud rotates around an axis parallel to the mountain range.

Runway Visual Range (RVR). An instrumentally-derived value, based on standard calibrations, that represents the horizontal distance a pilot may see down the runway from the approach end.

sand. Loose particles of granular material.

sandstorm. Particles of sand carried aloft by a strong wind. The sand particles are mostly confined to the lowest ten feet, and rarely rise more than fifty feet above the ground.

scattered. A layer whose summation amount of sky cover is 3/8ths through 4/8ths.

scheduled time of report. The time a scheduled report is required to be available for transmission.

sea-level pressure. The pressure value obtained by the theoretical reduction or increase of barometric pressure to sea-level.

sector visibility. The visibility in a specified direction that represents at least a 45 degree arc of the horizon circle.

shall. A term used to indicate that a standard is mandatory.

shallow. A descriptor, MI, used only to describe fog when the visibility at 6 feet above the ground is 5/8ths statute mile or more and the apparent visibility in the fog layer is less than 5/8ths statute mile.

shallow fog. Fog in which the visibility at 6 feet above ground level is 5/8ths statute mile or more and the apparent visibility in the fog layer is less than 5/8ths statute mile.

sheet ice. Ice formed by the freezing of liquid precipitation or the freezing of melted solid precipitation (see snow depth).

short-term storage. Storage of data for 4 or more days to assist in sensor/system maintenance and verification of sensor/system records in the event of an aircraft mishap.

should. A term used to indicate that a standard is recommended.

shower(s). A descriptor, SH, used to qualify precipitation characterized by the suddenness with which they start and stop, by the rapid changes of intensity, and usually by rapid changes in the appearance of the sky.

significant clouds. Cumulonimbus, cumulonimbus mammatus, towering cumulus, altocumulus castellanus, and standing lenticular or rotor clouds.

sky condition. The state of the sky in terms of such parameters as sky cover, layers and associated heights, ceiling, and cloud types.

sky cover. The amount of the sky which is covered by clouds or obscurations in contact with the surface.

small hail. See snow pellets.

smoke. A suspension in the air of small particles produced by combustion. A transition to haze may occur when smoke particles have traveled great distances (25 to 100 statute miles or more) and when the larger particles have settled out and the remaining particles have become widely scattered through the atmosphere.

snow. Precipitation of snow crystals, mostly branched in the form of six-pointed stars; for automated stations, any form of frozen precipitation other than hail.

snow depth. The vertical height of frozen precipitation on the ground. For this purpose, frozen precipitation includes ice pellets, glaze, hail, any combination of these, and sheet ice formed directly or indirectly from precipitation.

snow grains. Precipitation of very small, white, opaque grains of ice.

snow pellets. Precipitation of white, opaque grains of ice. The grains are round or sometimes conical. Diameters range from about 0.08 to 0.2 inch (2 to 5 mm).

spray. An ensemble of water droplets torn by the wind from an extensive body of water, generally from the crests of waves, and carried up into the air in such quantities that it reduces the horizontal visibility.

SPECI. A surface weather report taken to record a change in weather conditions that meets specified criteria or is otherwise considered to be significant.

squall. A strong wind characterized by a sudden onset in which the wind speed increases at least 16 knots and is sustained at 22 knots or more for at least one minute.

Standard Atmosphere. A hypothetical vertical distribution of the atmospheric temperature, pressure, and density, which by international agreement is considered to be representative of the atmosphere for pressure-altimeter calibrations and other purposes (29.92INS or 1013hPa).

standing lenticular cloud. A, more or less, isolated cloud with sharp outlines that is generally in the form of a smooth lens or almond. These clouds often form on the lee side of and generally parallel to mountain ranges. Depending on their height above the surface, they may be reported as stratocumulus standing lenticular cloud (SCSL); altocumulus standing lenticular cloud (ACSL); or cirrocumulus standing lenticular cloud (CCSL).

station elevation. The officially designated height above sea-level to which station pressure pertains. It is generally the same as field elevation at an airport station.

station identifier. A four alphabetic character code group used to identify the observing location.

Station Information File. A record that documents the site characteristics of an observing location and the reporting program at the location.

station pressure. The atmospheric pressure at the designated station elevation.

summation layer amount. A categorization of the amount of sky cover at and below each reported layer.

summation principle. This principle states that the sky cover at any level is equal to the summation of the sky cover of the lowest layer plus the additional sky cover provided at all successively higher layers up to and including the layer in question.

surface. The horizontal plane whose elevation above sea level equals the field elevation. At stations where the field elevation has not been established, the surface refers to the ground elevation at the observation site.

surface visibility. The prevailing visibility determined from the usual point of observation.

synoptic surface weather observation. Surface weather observations evaluated in accordance with WMO regulations (perhaps modified by national practices). These observations are reported no more frequently than every 3 hours.

temperature. A measure of the hotness or coldness of the ambient air as measured by a suitable instrument.

thunderstorm. A cumulonimbus cloud that is accompanied by lightning and thunder, or for automated systems, a storm detected by lightning detection systems.

time of occurrence. A report of the time weather begins and ends.

tornadic activity. The occurrence or disappearance of tornados, funnel clouds, or waterspouts.

tornado. A violent, rotating column of air touching the ground; funnel cloud that touches the ground (see funnel cloud and waterspout).

tower visibility. The prevailing visibility determined from the airport traffic control tower when the surface visibility is determined from another location.

towering cumulus. A descriptive term for a cloud with generally sharp outlines and with moderate to great vertical development, characterized by its cauliflower or tower appearance.

type of report. A code (METAR, SPECI) included in the weather report to indicate the content of the observation, and to indicate whether certain reporting criteria have been met.

type of station. A code figure (AO1, or AO2) for automated stations which is included in the remarks section of the report to indicate the scope of the observation program at the station that generated the report.

unknown precipitation. Precipitation type that is reported if the automated station detects the occurrence of precipitation but the precipitation discriminator cannot recognize the type.

variable ceiling. A ceiling of less than 3,000 feet which rapidly increases or decreases in height by established criteria during the period of observation.

variable layer amounts. A condition when the reportable amount of a layer varies by one or more reportable values during the period it is being evaluated (variable sky condition).

variable prevailing visibility. A condition when the prevailing visibility is less than 3 statute miles and rapidly increases and decreases by 1/2 mile or more during the period of observation.

variable wind direction. A condition when (1) the wind direction fluctuates by 60 degrees or more during the 2-minute evaluation period and the wind speed is greater than 6 knots; or (2) the direction is variable and the wind speed is 6 knots or less.

vertical visibility. A subjective or instrumental evaluation of the vertical distance into a surface-based obscuration that an observer would be able to see.

vicinity. A proximity qualifier, VC, used to indicate weather phenomena observed between 5 and 10 statute miles of the usual point of observation but not at the station.

virga. Visible wisps or strands of precipitation falling from clouds that evaporate before reaching the surface.

visibility. The greatest horizontal distance at which selected objects can be seen and identified or its equivalent derived from instrumental measurements.

visibility reference points. Selected objects at known distances from the weather station that are used to manually evaluate visibility.

volcanic ash. Fine particles of rock powder that originate from a volcano and that may remain suspended in the atmosphere for long periods.

volcanic eruption. An explosion caused by the intense heating of subterranean rock which expels lava, steam, ashes, etc., through vents in the earth's crust.

water equivalent. The liquid content of solid precipitation that has accumulated on the ground (snow depth). The accumulation may consist of snow, ice formed by freezing precipitation, freezing liquid precipitation, or ice formed by the refreezing of melted snow.

waterspout. A violent, rotating column of air that forms over a body of water, and touches the water surface; tornado or funnel cloud that touches a body of water (see funnel cloud and tornado).

well-developed dust/sand whirl. An ensemble of particles of dust or sand, sometimes accompanied by small litter, raised from the ground in the form of a whirling column of varying height with a small diameter and an approximately vertical axis.

weather. A category of individual and combined atmospheric phenomena which must be drawn upon to describe the local atmospheric conditions at the time of observation.

widespread dust. Fine particles of earth or other matter raised or suspended in the air by the wind that may have occurred at or far away from the station.

will. A term used to indicate futurity; it is not a requirement to be applied to standards.

wind. The horizontal motion of the air past a given point.

wind character. The description of the variability of the wind speed in terms of gusts.

wind direction. The true direction from which the wind is moving at a given location.

wind gust. See "gust."

wind shift. A change in the wind direction of 45 degrees or more in less than 15 minutes with sustained wind speeds of 10 knots or more throughout the wind shift.

wind speed. The rate at which air is moving horizontally past a given point. It may be a 2-minute average speed (reported as wind speed) or an instantaneous speed (reported as a peak wind speed, or gust).

APPENDIX B

LIST OF ABBREVIATIONS AND ACRONYMS[1]

---

[1]The abbreviations, acronyms, contractions, and symbols included in this appendix are defined in accordance with their usage in this Handbook.

$	maintenance check indicator
−	light intensity
+	heavy intensity
/	indicator that visual range data follows; separator between temperature and dew point data.
ACC	altocumulus castellanus
ACFT MSHP	aircraft mishap
ACSL	altocumulus standing lenticular cloud
ALP	airport location point
AO1	automated station without precipitation discriminator
AO2	automated station with precipitation discriminator
APRNT	apparent
APRX	approximately
ATCT	airport traffic control tower
AUTO	automated report
B	began
BC	patches
BKN	broken
BL	blowing
BR	mist
C	center (with reference to runway designation)
CA	cloud-air lightning
CB	cumulonimbus cloud
CBMAM	cumulonimbus mammatus cloud
CC	cloud-cloud lightning
CCSL	cirrocumulus standing lenticular cloud
CG	cloud-ground lightning
CHI	cloud-height indicator
CHINO	sky condition at secondary location not available
CIG	ceiling
CLR	clear
CONS	continuous

COR	correction to a previously disseminated report
DOC	Department of Commerce
DOD	Department of Defense
DOT	Department of Transportation
DR	low drifting
DS	duststorm
DSNT	distant
DU	widespread dust
DZ	drizzle
E	east, ended
FAA	Federal Aviation Administration
FC	funnel cloud
FEW	few clouds
FG	fog
FIBI	filed but impracticable to transmit
FIRST	first observation after a break in coverage at manual station
FMH-1	Federal Meteorological Handbook No.1, *Surface Weather Observations & Reports*
FMH-2	Federal Meteorological Handbook No.2, *Surface Synoptic Codes*
FROPA	frontal passage
FRQ	frequent
FT	feet
FU	smoke
FZ	freezing
FZRANO	freezing rain sensor not available
G	gust
GR	hail
GS	small hail and/or snow pellets
HZ	haze
IC	ice crystals, in-cloud lightning
ICAO	International Civil Aviation Organization
KT	knots
L	left (with reference to runway designation)

LAST	last observation before a break in coverage at a manual station
LST	Local Standard Time
LTG	lightning
LWR	lower
M	minus, less than
METAR	aviation routine weather report
MI	shallow
MOV	moved/moving/movement
MT	mountains
N	north
N/A	not applicable
NCDC	National Climatic Data Center
NE	northeast
NOS	National Ocean Service
NOSPECI	no SPECI reports are taken at the station
NW	northwest
NWS	National Weather Service
OCNL	occasional
OFCM	Office of the Federal Coordinator for Meteorology
OVC	overcast
OHD	overhead
P	greater than
PL	ice pellets
PK WND	peak wind
PNO	precipitation amount not available
PO	dust/sand whirls (dust devils)
PR	partial
PRESFR	pressure falling rapidly
PRESRR	pressure rising rapidly
PWINO	precipitation identifier sensor not available
PY	spray
R	right (with reference to runway designation)

RA	rain
RVR	Runway Visual Range
RVRNO	RVR system not available
RWY	runway
S	south
SA	sand
SCSL	stratocumulus standing lenticular cloud
SCT	scattered
SE	southeast
SFC	surface
SG	snow grains
SH	shower(s)
SKC	sky clear
SLP	sea-level pressure
SLPNO	sea-level pressure not available
SM	statute miles
SN	snow
SNINCR	snow increasing rapidly
SPECI	an unscheduled report taken when certain criteria have been met
SQ	squalls
SS	sandstorm
SW	southwest
TCU	towering cumulus
TS	thunderstorm
TSNO	thunderstorm information not available
TWR	tower
UP	unknown precipitation
UTC	Coordinated Universal Time
V	variable
VA	volcanic ash
VC	in the vicinity
VIS	visibility

VISNO	visibility at secondary location not available
VRB	variable
VV	vertical visibility
W	west
WG/SO	Working Group for Surface Observations
WMO	World Meteorological Organization
WND	wind
WSHFT	wind shift
Z	zulu, i.e., Coordinated Universal Time

APPENDIX C

SENSOR STANDARDS

SENSOR STANDARDS

C.1 Runway Visual Range Standards. Table C-1 lists the required accuracy in feet for runway visual range sensors.

Table C-1. Accuracy for Runway Visual Range

Runway Visual Range In Feet	Accuracy
<1,300 feet	±30 feet
1,300 through 2,600 feet	±80 feet
>2,600 feet	±10%

C.2 Visibility Sensor Standards. Table C-2 lists the accuracy for automated visibility sensors.

Table C-2. Accuracy of Automated Visibility Sensors

Visibility from Standard Visibility Sensor	Percentage of Data Within or Exceeding Given Range		
	At least 80% Within	No more than 18% Exceed	No more than 2% Exceed
0 through 1 1/4	± 1/4	± 1/2	± 1
1 1/2 through 1 3/4	+ 1/4, − 1/2	+ 1/2, − 3/4	± 1
2 through 2 1/2	± 1/2	± 1	± 1
3	+ 1/2, − 1	± 1	± 1
4 through 10	± 1 RV*	± 2 RV*	± 2 RV*
*RV = Reportable value, all other values in miles.			

C.3 Wind Sensor Standards. Table C-3 lists the units of measure, the range, the accuracy, and the resolution for pressure parameters.

Table C-3. Units of Measure, Range, Accuracy and Resolution of Wind Parameters

Parameter	Units of Measure	Range	Accuracy	Resolution
Direction	Degrees	1° to 360°	±5° when speed is ≥5 knots	10°
Speed	Knots	2 to 90 knots	±1 knot up to 10 knots ±10% above 10 knots	1 knot

C.4 Sky Condition Standards. Table C-4 lists the range and accuracy for sky condition parameters.

Table C-4. Range and Accuracy of Sky Condition Parameters

Parameter	Range	Accuracy	
		Amounts	Heights
Sky Condition	0 - 12,000 ft. (minimum)	±1 Reportable Value	±3 Reportable Values

C.5 Temperature and Dew Point Parameter Standards. Table C-5 lists the range, accuracy, and resolution in degrees Celsius for temperature and dew point parameters.

Table C-5. Temperature and Dew Point Sensor Accuracy and Resolution (C°)

Parameter	Range	Accuracy	Resolution
Temperature	−62 to −50	±1.1	0.1
	−50 to +50	±0.6	0.1
	+50 to +54	±1.1	0.1
Dew point	−34 to −24	±2.2	0.1
	−24 to −01	±1.7	0.1
	−01 to +30	±1.1	0.1

C.6 Pressure Parameter Standards.
Table C-6 lists the units of measure, the range, the accuracy, and the resolution for pressure parameters.

Table C-6. Units of Measure, Range, Accuracy and Resolution of Pressure Parameters

Parameter	Units of Measure	Range	Accuracy	Resolution
Station Pressure	Inches of Mercury	4	±0.02	0.005 inch
Altimeter Setting	Inches of Mercury	4	±0.02	0.01 inch
Sea-Level Pressure	Hectopascals	136	±0.68	0.1 hectopascal

APPENDIX D

RUNWAY VISUAL RANGE TABLES

RUNWAY VISUAL RANGE TABLES

Table D-1. RVR Transmittance Conversion Table for Tasker 400 and Equivalent Systems with 250-Foot Baseline -- Contrast Threshold 5.5 Percent

	DAY	NIGHT	
RVR (Ft)	LS 5	LS 5	RVR (Ft)
400			400
---------------	.0299	.0013	---------------
600			600
---------------	.1038	.0113	---------------
800			800
---------------	.1974	.0351	---------------
1000			1000
---------------	.2905	.0707	---------------
1200			1200
---------------	.3746	.1134	---------------
1400			1400
---------------	.4479	.1590	---------------
1600			1600
---------------	.5107	.2048	---------------
1800			1800
---------------	.5644	.2492	---------------
2000			2000
---------------	.6104	.2913	---------------
2200			2200
---------------	.6499	.3307	---------------
2400			2400
---------------	.6840	.3674	---------------
2600			2600
---------------	.7136	.4014	---------------
2800			2800
---------------	.7395	.4328	---------------
3000			3000
---------------	.7774	.4820	---------------
3500			3500
---------------	.8194	.5415	---------------
4000			4000
---------------	.8431	.5906	---------------
4500			4500
---------------	.8584	.6317	---------------
5000			5000
---------------	.8710	.6662	---------------
5500			5500
---------------	.8815	.6957	---------------
6000			6000
---------------	.8905	.7209	---------------
LS - Light Setting			

Note: When a given value of RVR is being reported, the transmittance shall be between the two adjacent values listed in the table.

Table D-2. RVR Transmittance Conversion Table for Tasker 500 and Equivalent Systems with 250-Foot Baseline -- Contrast Threshold 5.0 Percent

	DAY	NIGHT	
RVR (Ft)	LS 5	LS 5	RVR (Ft)
500			500
-----------------	.0449	.0027	-----------------
600			600
-----------------	.0823	.0075	-----------------
700			700
-----------------	.1264	.0159	-----------------
800			800
-----------------	.1974	.0351	-----------------
1000			1000
-----------------	.2905	.0707	-----------------
1200			1200
-----------------	.3746	.1134	-----------------
1400			1400
-----------------	.4479	.1590	-----------------
1600			1600
-----------------	.5107	.2048	-----------------
1800			1800
-----------------	.5644	.2492	-----------------
2000			2000
-----------------	.6104	.2913	-----------------
2200			2200
-----------------	.6499	.3307	-----------------
2400			2400
-----------------	.6840	.3674	-----------------
2600			2600
-----------------	.7136	.4014	-----------------
2800			2800
-----------------	.7395	.4328	-----------------
3000			3000
-----------------	.7774	.4820	-----------------
3500			3500
-----------------	.8190	.5415	-----------------
4000			4000
-----------------	.8384	.5906	-----------------
4500			4500
-----------------	.8541	.6317	-----------------
5000			5000
-----------------	.8671	.6662	-----------------
5500			5500
-----------------	.8779	.6957	-----------------
6000			6000
-----------------	.8871	.7209	-----------------
LS - Light Setting.			

* Nighttime readings on the recorder cannot be determined for less than 700 feet. Report, if appropriate, M0700FT.

Note: When a given value of RVR is being reported, the transmittance shall be between the two adjacent values listed in the table.